State-of-the-Art Renewable Energy in Korea

State-of-the-Art Renewable Energy in Korea

Editors

Zong Woo Geem
Junhee Hong
Woohyun Hwang

MDPI • Basel • Beijing • Wuhan • Barcelona • Belgrade • Manchester • Tokyo • Cluj • Tianjin

Editors
Zong Woo Geem
Department of Energy IT
Gachon University
Seongnam
Korea

Junhee Hong
Department of Energy IT
Gachon University
Seongnam
Korea

Woohyun Hwang
Head Office
Jeju Energy Corporation
Jeju
Korea

Editorial Office
MDPI
St. Alban-Anlage 66
4052 Basel, Switzerland

This is a reprint of articles from the Special Issue published online in the open access journal *Applied Sciences* (ISSN 2076-3417) (available at: https://www.mdpi.com/journal/applsci/special_issues/Renewable_Energy_Korea).

For citation purposes, cite each article independently as indicated on the article page online and as indicated below:

LastName, A.A.; LastName, B.B.; LastName, C.C. Article Title. *Journal Name* **Year**, *Volume Number*, Page Range.

ISBN 978-3-0365-5332-0 (Hbk)
ISBN 978-3-0365-5331-3 (PDF)

Contents

Editorial

Special Issue on State-of-the-Art Renewable Energy in Korea

Zong Woo Geem [1,*], Junhee Hong [1] and Woohyun Hwang [2]

1   Department of Energy IT, Gachon University, Seongnam 13120, Korea; hongpa@gachon.ac.kr
2   Jeju Energy Corporation, Jeju 63219, Korea; hwangwh@jejuenergy.or.kr
*   Correspondence: geem@gachon.ac.kr

Citation: Geem, Z.W.; Hong, J.; Hwang, W. Special Issue on State-of-the-Art Renewable Energy in Korea. *Appl. Sci.* **2021**, *11*, 4401. https://doi.org/10.3390/app11104401

Received: 7 May 2021
Accepted: 8 May 2021
Published: 12 May 2021

Publisher's Note: MDPI stays neutral with regard to jurisdictional claims in published maps and institutional affiliations.

Nowadays, renewable energy plays an important role in nationwide power systems. We previously dealt with the problem of accepting renewable energy; now we deal with utilizing it. This Special Issue addresses three major aspects of the current trend towards the use of renewable energy in South Korea.

The first aspect is a renewable-based power system, where both main and ancillary supplies are sourced from renewable energies. Ko et al. [1] proposed an incentive model for ESS (energy storage system) utilization in order to reduce the fluctuation of wind power. They applied it to Jeju island which has a very high proportion of renewable energy. Similarly, Lee and Kim [2] proposed an economic model for ESS-based frequency regulation from the electricity market price forecast in Korea. ESS has an advantage in terms of faster response to frequency variation than conventional fossil-fuel generators. Ko et al. [3] developed a demand-side management model using a demand response (DR) aggregator and showed real cases in South Korea. The paper analyzes the economic effect of the DR program.

The second aspect is a distribution network for renewable energy. Kim et al. [4] proposed an optimal operation scheduling model using an energy band in a microgrid. The model operates between a distribution network (DN) and microgrid (MG) while minimizing the cost of the DN and maximizing the profit of MG. A major issue of the DN is a scheme for coordination of the protection relays needed for fault currents. The model proposed by Wadood et al. [5] minimizes the total operating time of all relays to prevent excessive interruptions.

The final aspect is a nano grid network technology. Lee [6] and Shin and Geem [7] show examples of a house while Park and del Pobil [8] show a building. This is a meaningful and timely approach with respect to an ESG (Environment, Social, Governance) trend.

The papers compiled in this special issue do not suggest that the increase in renewable energy is simply the replacement of fossil energy. Renewable energy requires many innovations over existing power infrastructure and regulation. These articles show the changing trend in various sectors in Korea.

Acknowledgments: We thank all the authors, reviewers, and staffs (especially Nicole Lian) for their contributions to this special issue.

Conflicts of Interest: The authors declare no conflict of interest.

References

1. Ko, W.; Lee, J.; Kim, J. The Effect of a Renewable Energy Certificate Incentive on Mitigating Wind Power Fluctuations: A Case Study of Jeju Island. *Appl. Sci.* **2019**, *9*, 1647. [CrossRef]
2. Lee, E.; Kim, J. Assessing the Benefits of Battery Energy Storage Systems for Frequency Regulation, Based on Electricity Market Price Forecasting. *Appl. Sci.* **2019**, *9*, 2147. [CrossRef]
3. Ko, W.; Vettikalladi, H.; Song, S.-H.; Choi, H.-J. Implementation of a Demand-Side Management Solution for South Korea's Demand Response Program. *Appl. Sci.* **2020**, *10*, 1751. [CrossRef]
4. Kim, H.-Y.; Kim, M.-K.; Kim, H.-J. Optimal Operational Scheduling of Distribution Network with Microgrid via Bi-Level Optimization Model with Energy Band. *Appl. Sci.* **2019**, *9*, 4219. [CrossRef]
5. Wadood, A.; Gholami Farkoush, S.; Khurshaid, T.; Kim, C.-H.; Yu, J.; Geem, Z.W.; Rhee, S.-B. An Optimized Protection Coordination Scheme for the Optimal Coordination of Overcurrent Relays Using a Nature-Inspired Root Tree Algorithm. *Appl. Sci.* **2018**, *8*, 1664. [CrossRef]
6. Lee, H.-K. Designing a Waterless Toilet Prototype for Reusable Energy Using a User-Centered Approach and Interviews. *Appl. Sci.* **2019**, *9*, 919. [CrossRef]
7. Shin, H.; Geem, Z.W. Optimal Design of a Residential Photovoltaic Renewable System in South Korea. *Appl. Sci.* **2019**, *9*, 1138. [CrossRef]
8. Park, E.; Del Pobil, A.P. Eco-Friendly Education Facilities: The Case of a Public Education Building in South Korea. *Appl. Sci.* **2018**, *8*, 1733. [CrossRef]

 *applied*
sciences

Article

Assessing the Benefits of Battery Energy Storage Systems for Frequency Regulation, Based on Electricity Market Price Forecasting

Eunjung Lee and Jinho Kim *

School of Integrated Technology, Gwangju Institute of Science and Technology, 123 Cheomdangwagi-ro, Buk-gu, Gwangju 61005, Korea; jkl51149@gist.ac.kr
* Correspondence: jeikim@gist.ac.kr; Tel./Fax: +82-62-715-5322

Received: 24 April 2019; Accepted: 20 May 2019; Published: 26 May 2019

Featured Application: Authors are encouraged to provide a concise description of the specific application or a potential application of the work. This section is not mandatory.

Abstract: In electricity markets, energy storage systems (ESSs) have been widely used to regulate frequency in power system operations. Frequency regulation (F/R) relates to the short-term reserve power used to balance the real-time mismatch of supply and demand. Every alternating current power system has its own unique standard frequency level, and frequency variation occurs whenever there is a mismatch of supply and demand. To cope with frequency variation, generating units—particularly base-loader generators—reduce their power outputs to a certain level, and the reduced generation outputs are used as a generation reserve whenever frequency variation occurs in the power systems. ESSs have recently been implemented as an innovative means of providing the F/R reserve previously provided by base-loader generators, because they are much faster in responding to frequency variation than conventional generators. We assess the economic benefits of ESSs for F/R, based on a new forecast of long-term electricity market price and real power system operation characteristics. For this purpose, we present case studies with respect to the South Korean electricity market as well as simulation results featuring key variables, along with their implications vis-à-vis electricity market operations.

Keywords: frequency regulation; energy storage system; economic benefits; price forecast; electricity market operation

---

1. Introduction

Global electricity markets have started to use energy storage systems (ESSs) to enhance the operational performance efficiency of power systems. Compared to other existing resources such as coal and gas-fueled generators, ESSs respond to changes in demand much more quickly. This feature offers great operational flexibility in the electricity market and in system operations, particularly in the smart operation of frequency regulation (F/R). F/R is an activity through which system operations cope with excessive fluctuations in power system frequency—fluctuations that are caused by real-time mismatches in power supply and demand. Conventional coal and/or gas-type generators have been traditionally used to resolve the F/R problem, by leaving some portion of their generation capacity unused—that is, by procuring generation reserves, and by providing reserved resources in the event of excessive frequency fluctuation.

Given the technical advantage of ESSs in terms of their prompt responsiveness to frequency fluctuation, electricity markets in the United States—such as Pennsylvania-New Jersey-Maryland(PJM) Interconnection, Midcontinent Independent System Operator(MISO), and New York Independent System Operator(NYISO)—have already designed ESS practices in F/R markets, and they have attracted the entry of ESSs through multiple incentive mechanisms [1]. For example, some F/R markets have introduced an incentive mechanism divided into a capacity market and an energy market, to offer more benefits to those resources that respond accurately and rapidly [2–4]. Moreover, because power system frequency signals can be more frequently transmitted to ESS than conventional generators, PJM and NYISO operate single transmission systems that are separated into fast and slow-response resources [5,6]. In addition, ESSs are being more broadly applied to electricity systems: ESSs, for example, are typically associated with connecting variable renewable energy sources to enhance the power of battery charging, as well as to effectively operate and utilize electric vehicles; both are typical recent examples of ESS applications in power system operations [7–11]. In some electricity markets, including that of South Korea, electricity utility companies have undertaken large-scale ESS deployment plans for F/R.

To date, various studies have been conducted on F/R ESSs, covering topics such as optimal ESS capacity estimation and economic benefit assessments. Some studies [12–14] discuss the optimal capacity estimation of ESS for frequency control and evaluate the benefits thereof. Other studies [15,16] suggest the economic dispatch methodology and the optimal sizing of ESSs from a utility operation perspective. In determining benefits, an economic assessment should precede ESS installation. Hur et al. [17] propose economic analysis when an ESS is introduced as an F/R resource in an electricity market. Some studies discuss, from a utility perspective, economic benefit analyses in accordance with price arbitrage as a result of ESS application [18–20]. In economic analyses of electricity markets, the long-term system marginal price (SMP) estimation is the most important factor; however, most studies conduct short- and medium-term SMP estimation. Conejo et al. [21] conducted short-term SMP estimations using 24-h electricity price predictions for the day-ahead energy market, by applying various methodologies (i.e., neural network, time series, and wavelet models). Paraschiv et al. [22] propose a regime-switching model for short- and medium-term electricity price forecasting and show the superiority of the proposed model compared to an autoregressive integrated moving average and generalized autoregressive conditional heteroscedasticity models. Nowotarski et al. [23] discuss a long-term seasonal component that considers annual seasonality and estimates a future (one-year) electricity spot price by applying a wavelet-based model.

The current study proposes an analytical method by which to assess the benefit of ESS implementation for F/R in electricity markets. First, to capture the basic benefit of ESS for F/R, we developed a method of predicting the SMP, which is the weighted mean of the fuel cost of a marginal plant. Second, we proposed a new scenario-based method to forecast utility economic benefits; this method considers both the electricity market structure and power system operations. The case study results show the diverse profile of the economic aspects of ESS implementation; one can readily infer from the results economic insights pertaining to large-scale ESS implementation. This study contributes to the literature on economics analysis and long-term SMP estimation. this study contributes to the literature on two perspectives. First of all, a benefits analysis of ESS for F/R is conducted in accordance with electricity market in South Korea. And the proposed methodology, Long-term SMP estimation doesn't require large time series data, therefore there shouldn't be too much difficulty with respect to data collection.

The remainder of this paper is organized as follows. In Section 2, we develop a novel methodology for assessing the economic benefits of ESS implementation in the F/R electricity market, based on South Korea's national plan for long-term electricity supply and demand [24]. Section 3 addresses the simulation results by using the methods proposed in this study, while our conclusions are presented in Section 4.

2. Benefits Assessment of ESS Introduction in the F/R Market

2.1. ESS Introduction: Benefits Overview

The economic benefits of ESSs for F/R derive primarily from the difference in generation cost (i.e., fuel cost in \$/kWh) between base and peak-loader generators. To balance the mismatch in supply and demand in the real-time operation of a power system, a certain amount of a base-loader generator's capacity (typically 5%) is reserved for power system F/R. Instead, to meet load demands, expensive peak-loader generators produce the required electric power not otherwise supplied by base load generators. In this way, use of this F/R reserve causes an increase in the power system operational fuel cost.

However, the reserved generation amount offered by base-loader generators can be replaced by introducing an ESS for F/R. In other words, the reserved amount from base-loader generators—which are cheaper than peak-loaders—can be supplied to power systems to meet load demand. From the standpoint of power system operation, the use of ESSs for F/R facilitates the replacement of expensive peak-loader generators with cheap base-loaders.

Figure 1 depicts the basic economic benefit of an ESS for F/R in the electricity market. One of the key factors in assessing the economic benefit of ESSs for F/R in power system operation is the estimation of future prospects for the SMP in electricity markets. The SMP is the spot market price used in electric power transactions, and it is determined by considering the most expensive generation cost of the marginal generating unit that meets the marginal demand of electricity markets. When it comes to a base load generator's reserved generation associated with F/R, the revenue lost by not selling the reserved generation can be compensated for by offering the opportunity cost (COFF), which is defined as the difference between the SMP and the base-loader generator's fuel cost. Because the SMP, or the generation cost of a marginal generator, is typically decided by the peak load generator's fuel cost, the COFF offered to base load generators for F/R can be redefined as the difference between the generation costs of peak and base-loaders. In this regard, the economic benefit of ESSs for F/R can be captured by the replacement benefit—that is, the benefit that derives from fuel cost savings on account of replacing expensive peak-loader generators with cheaper base loaders in F/R. The benefit can, therefore, be assessed primarily by forecasting future SMP (i.e., the generation cost of peak load) and base load fuel costs.

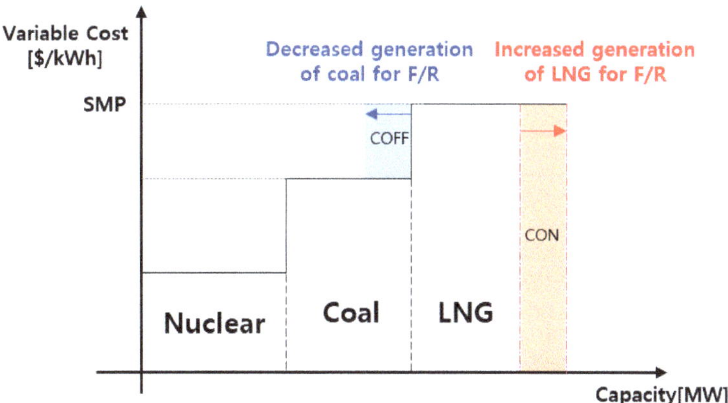

Figure 1. Economic benefits of ESSs for F/R in the electricity market [25,26].

2.2. Probabilistic Long-Term SMP Forecast

To assess effectively the economic benefit of ESS for F/R over a given horizon (typically 10–15 years), we propose the novel probabilistic weighted average to predict future annual SMP profiles for the horizon. Because the SMP is the most expensive fuel cost of the generator that is last committed to meet the forecasted demand at a given hour, the estimation for an hourly marginal plant profile across the operating horizon is the key element in predicting annual SMPs. To this end, the current study proposes a probabilistic method by which to forecast a long-term marginal plant profile associated with hourly SMP, to assess the economic benefit of ESS for F/R.

To obtain the annual SMP profiles—that is, annual marginal plant probability profiles in electricity markets—we used a 15-year national long-term supply and demand projection plan published by a South Korean energy agency [24]. Detailed descriptions of the development of a probabilistic annual SMP forecast, based on the estimation of a long-term marginal plant profile, are given below.

First, the annual generation capacity of each fuel type for the next 10 years can be obtained from the national plan for long-term electricity supply and demand. However, this capacity cannot be identified as real generation capacity, because it does not take into account the operational unavailability of generators owing to events such as forced and maintenance outages. The forced outage rate (FOR) and maintenance outage rate (MOR) speak to the unavailability of generating units associated with unplanned and planned outages, respectively. When assessing the annual generation capacity for each fuel type, the unavailable generation capacity should therefore be extracted from the nominal generation capacity. In addition, given the fluctuating output of renewable energy, we use estimations of actual generation capacity from the national plan, rather than installed capacity data.

Second, we obtain from the hourly demand average the past demand profiles that are assumed to be identical to estimated peak demand. However, these profiles can also be divided into two different demand profiles—weekdays and weekends. Annual peak demand for the next 10 years is used to determine the annual hourly demand. The methodology is as follows. Equations (1) and (2) represent the idea that the sum of the hourly average demand is the product of hourly peak demand and α_t (rate of hourly demand on the basis of peak demand), which transform to Equation (3). Applying this notion, annual peak demand satisfying average demand is calculated by Equation (4). Peak demand is calculated as:

$$d_t = d_{peak} \times \alpha_t \tag{1}$$

$$\sum_{t=1}^{24} d_t = \sum_{t=1}^{24} d_{peak} \times \alpha_t = d_{peak} \sum_{t=1}^{24} \alpha_t \tag{2}$$

$$d_{peak} \sum_{t=1}^{24} \frac{\alpha_t}{24} = d_{average} \tag{3}$$

$$d_{peak} = \frac{d_{average}}{\sum_{t=1}^{24} \alpha_t} \tag{4}$$

where $d_t (0 \le t \le 24)$, $d_{average}$, and $\alpha_t (0 \le \alpha_t \le 1)$ are the demand at each time, peak demand, and rate of past demand for each time on the basis of peak demand, respectively.

Third, the annual demand clustering pattern can be obtained from the peak demand for each year, as drawn from past data. This means that estimated demand is equal to the movement of the past demand pattern, in line with peak demand. Therefore, the annual demand pattern is estimated by multiplying annual peak demand by each value of the percentage of demand for every hour, based on peak demand from the past demand clustering pattern. We compare the annual generation mix from generation capacity and the demand clustering for every hour to identify the marginal plant resources used on weekdays, weeknights, and weekends. Marginal plant profiles for daytime, nighttime, and weekends are realized by designating daytime as 16 h, nighttime as 8 h (i.e., 12 AM–8 AM), and weekends as 24 h. This can be used to count numbers determined as SMP for the specific resource.

Fourth, this study assigns weighting for generation costs, such that they are allocated a heavier weight when they are closer to the present; it is assumed that future generation costs will be similar to past generation costs. The SMP for weekdays, weeknights, and weekends is estimated by using the weighted average of the marginal plant profile. Future SMP is calculated as follows.

$$SMP = C_{nuclear} \times P(X_{nuclear}) + C_{coal} \times P(X_{coal}) + C_{LNG} \times P(X_{LNG}) + C_{oil} \times P(X_{oil}) \tag{5}$$

where the subscripts $C_{nuclear}$, C_{coal} C_{LNG}, and C_{oil} denote nuclear power, coal, liquefied natural gas (LNG), and oil generation costs, respectively; subscripts $X_{nuclear}$, X_{coal}, X_{LNG}, and X_{oil} denote nuclear power, coal, LNG, and oil variables, respectively; and $P(X_i)$ denotes the marginal plant profile of X_i.

Using this function, the SMP for daytime, nighttime, and weekends can be determined. The annual SMP contains the rates for daytime (approximately 16 h per day for five days per week), nighttime (approximately 8 h per day for five days per week), and weekend (24 h per day for two days per week). In accordance with supposition, the outcomes of rate calculation are 0.476, 0.238, and 0.286, respectively, on the basis of one year (8760 h); we assign these rates as a calculus in Equation (6). The annual SMP associated with these rates is defined as

$$SMP_A = SMP_d \times 0.476 + SMP_n \times 0.238 + SMP_w \times 0.286. \tag{6}$$

SMP_A, SMP_d, SMP_n, and SMP_w are annual, daytime, nighttime, and weekend SMPs, respectively. The SMP_A formula consists of SMP_d, SMP_n, and SMP_w, with their weights calculated by using the duration rate in the year. The long-term SMP estimation framework is illustrated as Figure 2.

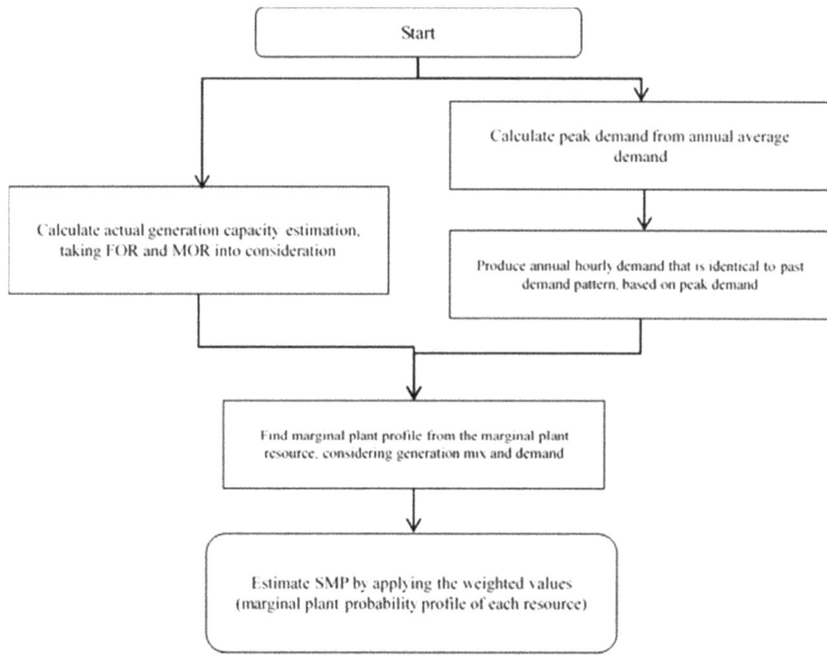

Figure 2. Schematic diagram of the SMP forecast methodology.

2.3. Assessment of Economic Benefits from ESS for F/R in the Electricity Market

ESS is introduced in a power system for F/R. If implemented in the South Korean electricity market, it will change the overall demand placed on coal and LNG supply capacity generators and modify electricity costs.

Currently, coal generators need to secure a reserve for F/R through a 5% reduced operation in the electricity market. Although this generation constraint is not included in the price-setting schedule, it is used in the operation schedule that is produced following the creation of the price-setting schedule. Therefore, reduced coal capacities receive an opportunity cost payment known as a constrained-off energy payment (i.e., the aforementioned COFF), which is calculated based on the minimum SMP and coal fuel cost. To meet the shortfall, LNG generators increase generation and then receive compensation, known as a constrained-on energy payment (CON); CON is calculated based on the maximum SMP and LNG fuel cost. However, the settlement would differ when operating ESS for F/R: because F/R ESSs can alleviate the constraint, coal capacities can generate more, and be compensated for this increased generation in the form of a scheduled energy payment (SEP). This SEP is calculated based on the SMP in the price-setting schedule, rather than the COFF. Furthermore, the LNG generators need not generate more to compensate for the shortfall, and so they do not receive the CON payment. As a result, each participant (i.e., Coal Gen., LNG Gen., and Utility) would then be compensated as in Table 1.

Table 1. Payment changes deriving from frequency regulation energy storage systems introduction to the South Korean energy market.

Participants	Before	After	Benefits
Coal Gen.	COFF	SEP	SEP–COFF
LNG Gen.	CON	–	–CON
Utility	CON+COFF	SEP	SEP–CON–COFF

We propose a method by which to estimate the utility benefits (UBs) of introducing into electricity markets ESSs for F/R. We consider the benefits in the energy and ancillary service (A/S) markets, based on generation constraints and when considering F/R in the South Korean electricity market.

Given that ESSs can be used as reserves, the implementation of an ESS alleviates the base load generation constraint and can produce benefits similar to ESS capacity; this is because the utility need not pay additional costs with respect to the coal and LNG generators. In addition, an A/S payment would be provided to the utility because the F/R ESS, which the utility needs to plan to own, replaces the conventional generation role. Therefore, the UBs increase in terms of the energy and A/S aspects. Equations for calculating the energy market price (EP) and the A/S price (ASP) benefits are as follows.

$$EP = Availability_{ESS} \times (CON + COFF - SEP) \times 8760 \times R_{Power} \tag{7}$$

where $Availability_{ESS}$ and R_{Power} are the coefficient of utilization for ESS and the generation operation rate, respectively, and

$$ASP = Capacity_{ESS} \times Availability_{ESS} \times W_{ESS} \times W_{Droop} \times W_{Deadband} \times U_{FR} \times R_{ESS} \times 8760 \tag{8}$$

where $Capacity_{ESS}$, W_{ESS}, W_{Droop}, and $W_{Deadband}$ are the practical ESS capacity, weighted values of ESS, droop, and dead band, respectively. Furthermore, U_{FR} is the unit cost for F/R, and R_{ESS} is the ESS operation rate. ESS compensation should be differentiated from conventional resource compensation, because it provides outstanding F/R performance; therefore, resource weighting was added through W_{ESS}—which has a value exceeding 1 in the ESS settlement of the A/S market—to provide a larger payment than conventional resources. Both the droop and dead band demonstrate the performance of resources in F/R, and thus, these factors should also be considered in the A/S settlement by using W_{Droop} and $W_{Deadband}$. These have values in the range of 0.8–1.05 and 0.85–1.05, respectively, and resources with lower values in them are set so as to have heavy weighting. The sum of energy (EP) and A/S benefits (ASP) equals the UB, given by Equation (9).

$$UB = EP + ASP \tag{9}$$

Although we actually assume that UBs reflecting the current electricity market need to come about during the daytime of a weekday, we also consider two other cases to make a total of three. In the first case, UBs occur during the daytime of a weekday, because coal generators generate more electricity during the day than at other times. In the second, UBs are derived during the nighttime of a weekday, because the upward generation of coal generators deepens during that time. In the third, UBs come about all day, on account of a stable trend of reserves and little upward generation among coal generators. For each of these cases, we present below equations by which to calculate EP and ASP benefits.

1. Case A: benefits occur during the daytime

$$EP = Availability_{ESS} \times (CON + COFF - SEP) \times 8760 \times 47.6\% \tag{10}$$

$$ASP = Capacity_{ESS} \times Availability_{ESS} \times W_{ESS} \times W_{Droop} \times W_{Deadband} \times U_{FR} \times R_{ESS} \times 8760 \times 66.7\% \tag{11}$$

2. Case B: benefits occur during the nighttime

$$EP = Availability_{ESS} \times (CON + COFF - SEP) \times 8760 \times 28.8\% \tag{12}$$

$$ASP = Capacity_{ESS} \times Availability_{ESS} \times W_{ESS} \times W_{Droop} \times W_{Deadband} \times U_{FR} \times R_{ESS} \times 8760 \times 33.3\% \tag{13}$$

3. Case C: benefits occur all day

$$EP = Availability_{ESS} \times (CON + COFF - SEP) \times 8760 \times 100\% \tag{14}$$

$$ASP = Capacity_{ESS} \times Availability_{ESS} \times W_{ESS} \times W_{Droop} \times W_{Deadband} \times U_{FR} \times R_{ESS} \times 8760 \times 100\% \tag{15}$$

3. Case Study: ESS Participation Benefits Assessment in the F/R Market, Based on the Long-Term SMP Forecasting Methodology

3.1. Comparison of Real and Estimated SMPs

We measure real SMP against estimated SMP and use past data from the Korean Power Exchange (KPX) to assess the results of our proposed methodology. To compare the SMP based on real data to the estimated SMP, we use real SMP for each month, forecasted SMP using marginal plant probability, and generation cost, using data from the 2001–2015 period. Figure 3 presents the results.

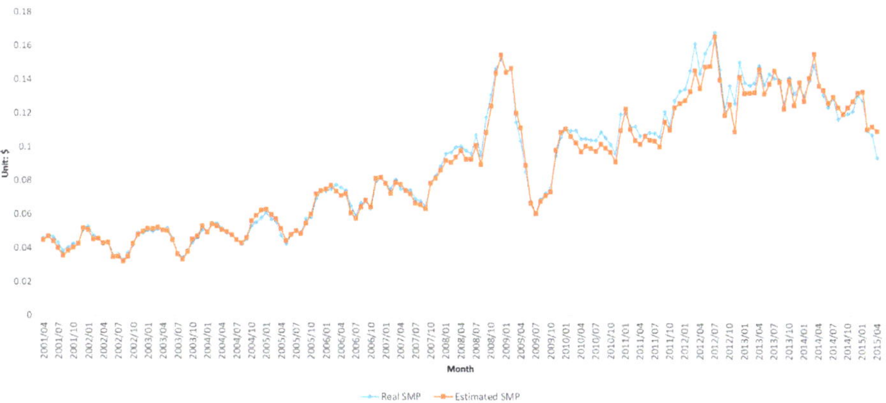

Figure 3. Comparison of real and estimated SMP, 2001–2015.

The results show that the estimations are similar to the real SMP values, with an average error of approximately 3.5%. Therefore, we determined that it is possible to predict the SMP by using the methodology presented herein.

3.2. Long-Term SMP Forecasting of the South Korean Electricity Market

We conducted a case study to investigate UBs and examine how ESS for F/R affects efficiency in South Korea. To undertake this investigation, we must first estimate the SMP according to the method proposed in Section 2.1. First, we considered the auxiliary power consumption ratio, the FOR, and the MOR to determine the actual supply capacity; this was obtained by deducting these rates for each of the generation resources. We assumed that the auxiliary power consumption ratio was 5%; additionally, for FOR and MOR, we used the average from the 2014–2015 period (Table 2).

Table 2. Average forced outage rate (FOR) and maintenance outage rate (MOR) values of South Korean nuclear and coal power plants.

Resources	Nuclear		Coal	
	FOR	MOR	FOR	MOR
2014	1.35%	15.69%	0.30%	7.56%
2015	1.96%	14.45%	0.30%	8.57%
Average	1.67%	15.07%	0.30%	8.06%

By substituting these figures and applying the renewable energy availability in resource capacity [24], we can determine the actual supply capacity and generation mix. The generation mix of the base load is based on the constructed actual supply capacity (Table 3). As part of the renewable energy penetration policy, the capacities of nuclear and coal resources will be reduced, while those of renewable energy resources will be increased. According to the national plan [24], renewable resources consist of photovoltaic power, wind power, tidal power, and by-product gas.

Table 3. Actual supply capacity of renewable, nuclear, and coal resources, 2019–2028 (Unit: MW).

Year	Renewable	Nuclear	Coal
2019	3704	20,712	34,121
2020	4045	20,712	35,305
2021	4398	20,712	37,796
2022	4756	21,825	39,813
2023	5117	22,422	39,813
2024	5799	21,666	38,752
2025	5799	21,666	38,752
2026	6691	18,844	37,805
2027	7191	17,532	37,805
2028	7699	16,561	34,559

Once we estimate the actual supply capacity, we can then estimate the annual demand on weekdays and weekends and compare these values to the actual supply capacity. Demand is estimated based on [24], and we use weekday and weekend patterns from 2016 to estimate the annual demand pattern ratio for 24 h; future demands are projected using the annual electricity consumption projection of [24], based on the load-pattern ratio. The hourly load profiles of weekdays and weekends, which are estimated as the average of the 2016 electricity load, are illustrated in Figure 4.

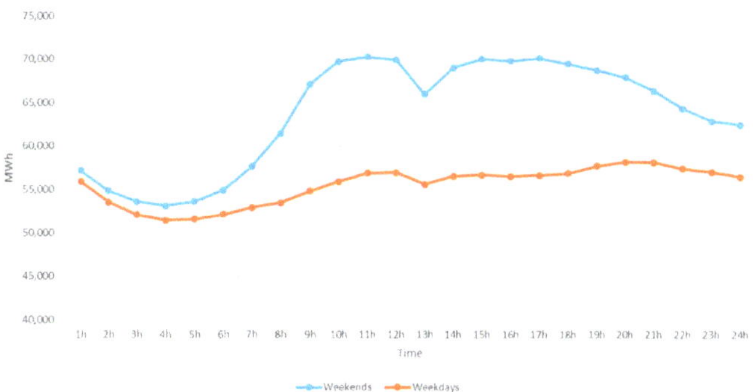

Figure 4. Hourly load profiles of weekday and weekend power usage.

We assume that the 2019–2028 load patterns are similar to the 2016 load patterns. To estimate the 2019–2028 load profile, the load-pattern ratio was obtained by dividing the load profile by the load profile peak demand. Annual peak load was calculated from the electricity consumption and the sum of the load-pattern ratio for 24 h (Table 4). We were then able to estimate the annual load profile of weekdays and weekends for the 2019–2028 period.

Table 4. Annual total consumption and peak load for 2019–2028, based on 2016 data.

Year	Electricity Consumption (GWh)	Peak Load (MW)
2019	537,973	88,172
2020	552,291	90,519
2021	566,714	92,883
2022	579,611	94,996
2023	592,145	97,051
2024	604,066	99,004
2025	615,788	100,926
2026	627,064	102,774
2027	637,866	104,544
2028	647,946	106,192

Following our 2019–2028 load profile estimation, we verified which resource would be selected in each hour by comparing the load profile and generation mix. Figure 5 shows the weekday and weekend load profiles, as well as generation, in 2024. Renewable resources do not comprise a single resource; rather, they are derived from multiple sources. Nonetheless, the national plan does not provide planned capacity for each type of renewable resource. This means that we encountered difficulties in generating actual hourly generation projections for renewable resources. Therefore, despite the inherent flexibility, we assumed that the single renewable resource is constant over time.

From these demand- and supply-side processes, we can obtain the marginal plant probability for each generation resource. Comparing the generation mix and demand allows us to see how frequently specific resources are selected as the marginal plant. Table 5 shows the marginal plant probability profile of coal generation.

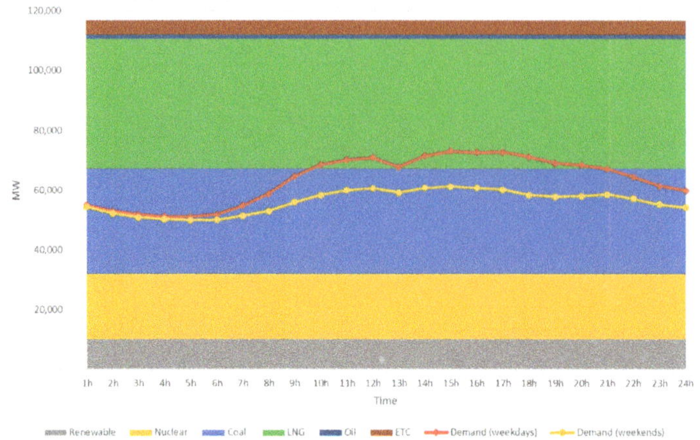

Figure 5. Hourly load profile of weekdays and weekends, and generation mix, in 2014.

Table 5. Marginal plant probability profile of coal generators, 2019–2028.

Year	Daytime	Nighttime	Weekends
2019	15.6%	62.0%	56.9%
2020	15.6%	62.0%	57.6%
2021	20.6%	64.8%	66.3%
2022	28.3%	71.3%	77.4%
2023	28.3%	71.3%	77.4%
2024	20.6%	64.8%	67.4%
2025	11.1%	57.4%	54.2%
2026	4.4%	49.1%	46.9%
2027	2.8%	45.4%	42.4%
2028	1.1%	44.4%	41.0%

We applied the average value of the annual generation cost from the 2016–2017 period. The SMP is estimated by using the marginal plant probability profile and generation cost and is shown in Table 6.

Table 6. System marginal price of coal generators, 2019–2028 (Unit: $).

Year	Daytime	Nighttime	Weekends
2019	7.20	5.47	6.29
2020	7.20	5.47	6.29
2021	7.01	5.36	6.08
2022	6.72	5.12	5.77
2023	6.72	5.12	5.77
2024	7.01	5.36	6.07
2025	7.36	5.64	6.44
2026	7.61	5.95	6.71
2027	7.67	6.09	6.83
2028	7.74	6.12	6.88

The SMP estimation results show a gradual decrease until 2023, and then a steady increase. The major driver of the SMP projection trend is recent energy policy that works to reduce South Korea's reliance on nuclear and coal power plants and expand its reliance on renewable resources.

3.3. Economic Assessment of ESS in F/R the Market, with Respect to UBs

Following the SMP estimation, we determined UBs as a function of introducing F/R ESS. The UBs consist of benefits in the energy and A/S markets. Simulations were conducted for each of case A, B, and C, using two ESS capacity scenarios that consider existing installed capacity (i.e., 52 MW) and planned future capacity (i.e., 500 MW). To calculate UBs in each scenario, the value of UBs is derived by using the following settings: $Availability_{ESS} = 48\%$, $W_{ESS} = 1.1$, $W_{Droop} = 1.05$, $W_{Deadband} = 1.05$ and $U_{FR} = 2.53$ \$/kWh. Furthermore, the daytime, nighttime, and annual average SMPs are applied to each equation in a regular sequence. Results pertaining to UBs during the 2019–2028 period are presented in Table 7.

Table 7. Projected Energy Market Price (EP) Benefits, Ancillary Service Price (ASP) Benefits, and Utility Benefits (UBs), 2019–2028 (Unit: Thousands of \$).

Case	EP Benefits		ASP Benefits		UBs	
	52 MW	500 MW	52 MW	500 MW	52 MW	500 MW
A	33,031	317,606	3192	30,689	36,223	348,295
B	7900	75,966	1596	15,345	9496	91,311
C	49,454	475,515	6703	64,448	56,157	539,963

In case A, over the 10-year period, for a 52-MW ESS (500-MW ESS), we could anticipate ASP benefits of about \$3 million (\$31 million), EP benefits of about \$33 million (\$318 million), and UBs of about \$36 million (\$348 million). When considering other cases based on case A, we can see that case B (case C) is 26% (155%) the scale of case A. In cases A, B, and C, the difference between the two capacity scenarios was approximately \$31, \$8, and \$48 million, respectively—demonstrating an 862% increase for an ESS capacity increase from 52 MW to 500 MW. This finding demonstrates that UBs are directly proportional to the size of the ESS. The benefit scale for these cases differs very much from the simulation results; this divergence derives from the fact that SMP change depending on the time involved and the times in which ESS benefits are generated. Consequently, changes made to the electricity industry—such as changes to energy policy, fuel costs, and demand among others—will decide the benefit level.

4. Conclusions

Recently, ESSs for electricity systems have been utilized in numerous ways. (For example, they are connected to renewable resources and used to discharge large quantities of electricity at peak usage times.) A plan to implement ESSs for F/R has recently been introduced in South Korea; other countries have already implemented them, because they allow for the stabilization of electricity systems in a way that compensates for its higher costs and encourages more efficient fuel use. We carried out a benefit estimation in anticipation of the introduction in South Korea of an F/R ESS.

We present a novel methodology for assessing the anticipated UBs. First, we extrapolated the future SMP by using a weighted average of marginal plant probability and fuel cost for each resource. We then calculated the UB as the sum of the energy and A/S market benefits, as determined by the electricity market and industrial structure. In the case study, we found the scale of benefits to range from \$91 million to \$540 million for 500 MW, and noted that among the three cases, case C—in which ESS for F/R is operated all day—offers the largest benefit. Although the results of the simulation models present different benefit levels, all cases show large and positive benefits; none show a negative result. Thus, we conclude that the introduction of ESSs for F/R in South Korea would enhance power system stability and bring about substantial UBs.

Author Contributions: Both authors contributed to this work. E.L. undertook related research, performed the analysis, and wrote the paper. J.K. designed the study and led and supervised the research.

Funding: This work was funded by the Korea Institute of Energy Technology Evaluation and Planning (KETEP) and the Ministry of Trade, Industry & Energy (MOTIE) of the Republic of Korea (grant number 20171210200810).

Conflicts of Interest: The authors declare no conflict of interest.

References

1. Wellinghoff, J. FERC Order No.755. Available online: https://www.ferc.gov/whats-new/comm-meet/2011/102011/E-28.pdf?csrt=2959623525451900142 (accessed on 22 October 2018).
2. Federal Energy Regulatory Commission (FERC). Order on Compliance Filing. Available online: http://energystorage.org/system/files/resources/ferc_orderapprovingmisos755compliancefiling_9_20_12.pdf (accessed on 21 April 2019).
3. Miller, R.; Venkatesh, B.; Cheng, D. Overview of FERC Order No.755 and Proposed MISO Implementation. In Proceedings of the 2013 IEEE Power & Energy Society General Meeting, Vancouver, BC, Canada, 21–25 July 2013.
4. Chen, Y.; Keyser, M.; Tackett, M.H.; Ma, X. Incorporating short-term stored energy resource into Midwest ISO energy and ancillary service market. *IEEE Trans. Power Syst.* **2011**, *26*, 829–838. [CrossRef]
5. KEMA. KERMIT Study Report. Available online: https://www.pjm.com/-/media/committees-groups/task-forces/rmistf/postings/pjm-kema-final-study-report.ashx?la=en (accessed on 22 April 2019).
6. PJM. RTO/ISO Regulation Market Comparison. Available online: https://www.pjm.com/-/media/committees-groups/task-forces/rmistf/20160413/20160413-item-03-rto-iso-benchmarking.ashx (accessed on 22 October 2018).
7. Dang, J.; Seuss, J.; Suneja, L.; Harley, R.G. SoC feedback control for wind and ESS hybrid power system frequency regulation. *IEEE J. Emerg. Sel. Top. Power Electron.* **2014**, *2*, 79–86. [CrossRef]
8. Datta, M.; Senjyu, T. Fuzzy control of distributed PV inverters/energy storage systems/electric vehicles for frequency regulation in a large power system. *IEEE Trans. Smart Grid* **2013**, *4*, 479–488. [CrossRef]
9. Zhou, H.; Bhattacharya, T.; Tran, D.; Siew, T.S.T.; Khambadkone, A.M. Composite energy storage system involving battery and ultracapacitor with dynamic energy management in microgrid applications. *IEEE Trans. Power Electr.* **2011**, *26*, 923–930. [CrossRef]
10. Wen, S.; Lan, H.; Fu, Q.; Yu, D.C.; Zhang, L. Economic allocation for energy storage system considering wind power distribution. *IEEE Trans. Power Syst.* **2015**, *30*, 644–652. [CrossRef]
11. Tran, D.; Khambadkone, A.M. Energy management for lifetime extension of energy storage system in micro-grid applications. *IEEE Trans. Smart Grid* **2013**, *4*, 1289–1296. [CrossRef]
12. Oudalov, A.; Chartouni, D.; Ohler, C. Optimizing a battery energy storage system for primary frequency control. *IEEE Trans. Power Syst.* **2007**, *22*, 1259–1266. [CrossRef]
13. Mercier, P.; Cherkaoui, R.; Oudalov, A. Optimizing a battery energy storage system for frequency control application in an isolated power system. *IEEE Trans. Power Syst.* **2009**, *24*, 1469–1477. [CrossRef]
14. Kottick, D.; Blau, M.; Edelstein, D. Battery energy storage for frequency regulation in an island power system. *IEEE Trans. Energy Convers.* **1993**, *8*, 455–459. [CrossRef]
15. Jung, K.-H.; Kim, H.; Rho, D. Determination of the installation site and optimal capacity of the battery energy storage system for load leveling. *IEEE Trans. Energy Convers.* **1996**, *11*, 162–167. [CrossRef]
16. Lo, C.H.; Anderson, M.D. Economic dispatch and optimal sizing of battery energy storage systems in utility load-leveling operations. *IEEE Trans. Energy Convers.* **1999**, *14*, 824–829. [CrossRef]
17. Hur, W.; Moon, Y.; Shin, K.; Kim, W.; Nam, S.; Park, K. Economic value of Li-ion energy storage system in frequency regulation application from utility firm's perspective in Korea. *Energies* **2015**, *8*, 5000–5017. [CrossRef]
18. Bradbury, K.; Pratson, L.; Patiño-Echeverri, D. Economic viability of energy storage systems based on price arbitrage potential in real-time U.S. electricity markets. *Appl. Energy* **2014**, *114*, 512–519. [CrossRef]
19. Walawalkar, R.; Apt, J.; Mancini, R. Economics of electric energy storage for energy arbitrage and regulation in New York. *Energy Policy* **2007**, *35*, 2558–2568. [CrossRef]
20. Alt, J.T.; Anderson, M.D.; Jungst, R.G. Assessment of utility side cost savings from battery energy storage. *IEEE Trans. Power Syst.* **1997**, *12*, 1112–1120. [CrossRef]
21. Conejo, A.J.; Contreras, J.; Espinola, R.; Plazas, M.A. Forecasting electricity prices for a day-ahead pool-based electric energy market. *Int. J. Forecast.* **2005**, *21*, 435–462. [CrossRef]

22. Paraschiv, F.; Fleten, S.E.; Schürle, M. A spot-forward model for electricity prices with regime shifts. *Energy Econ.* **2015**, *47*, 142–153. [CrossRef]

23. Nowotarski, J.; Tomczyk, J.; Weron, R. Robust estimation and forecasting of the long-term seasonal component of electricity spot prices. *Energy Econ.* **2013**, *39*, 13–27. [CrossRef]

24. The Eighth Basic Plan of Long-term Electricity Supply and Demand. Available online: https://www.kpx.or.kr/www/downloadBbsFile.do?atchmnflNo=28714 (accessed on 22 October 2018).

25. Electricity Market Operation Rule in South Korea. Available online: https://www.kpx.or.kr/www/selectBbsNttView.do?key=29&bbsNo=114&nttNo=18847&searchCtgry=&searchCnd=all&searchKrwd=&pageIndex=1&integrDeptCode= (accessed on 12 May 2019).

26. Korea Power Exchange. *Technical Report on ESS Operation System and Market Development for Frequency Regulation*; US Department of Energy: Washington, DC, USA, 2016; pp. 49–55.

 *applied sciences*

Article

Implementation of a Demand-Side Management Solution for South Korea's Demand Response Program

Wonsuk Ko [1], Hamsakutty Vettikalladi [1], Seung-Ho Song [2],* and Hyeong-Jin Choi [3]

[1] Department of Electrical Engineering, College of Engineering, King Saud University, Riyadh 11421, Saudi Arabia; wkoh@ksu.edu.sa (W.K.); hvettikalladi@ksu.edu.sa (H.V.)

[2] Department of Electrical Engineering, Kwangwoon University, Seoul 01897, Korea

[3] Department of Electrical Engineering, Kwangwoon University & GS E&C Corp, Seoul 01897, Korea; hjchoi@gsenc.com

* Correspondence: ssh@kw.ac.kr; Tel.: +82-2-940-5762; Fax: +82-2-940-5096

Received: 31 December 2019; Accepted: 26 February 2020; Published: 4 March 2020

Abstract: In this paper, we show the development of a demand-side management solution (DSMS) for demand response (DR) aggregator and actual demand response operation cases in South Korea. To show an experience, Korea's demand response market outline, functions of DSMS, real contracted capacity, and payment between consumer and load aggregator and DR operation cases are revealed. The DSMS computes the customer baseline load (CBL), relative root mean squared error (RRMSE), and payments of the customers in real time. The case of 10 MW contracted customers shows 108.03% delivery rate and a benefit of 854,900,394 KRW for two years. The results illustrate that an integrated demand-side management solution contributes by participating in a DR market and gives a benefit and satisfaction to the consumer.

Keywords: demand response; demand-side management solution; electricity market; energy management

---

1. Introduction

In a power system, electricity demand changes constantly. Power suppliers need to generate more power generation when demand is high, and less when demand is low. Traditionally, to coordinate supply and demand has been the supplier's responsibility and the demand side has been considered secondary. The power system supplier predicts the demand and, then, generates the supply capacity to meet the demand. After that, market price changed to supply capacity serves as criteria for deciding the facility capacity. The capacity that cannot be adjusted on the supply side is supplemented by DSM (demand-side management) such as temporarily reducing or moving the load on the demand side [1,2].

Recently, there has been a growing interest in considering the demand as the same as the supply side. Technological changes are occurring both on the supply side and on the demand side. Demand response is an alternative to additional infrastructure to maintain the safe margin between generation or distribution capacity and demand. The definition of demand response from the United State Department of Energy says, "Changes in electric usage by end-use customers from their normal consumption patterns in response to changes in the price of electricity over time, or to incentive payments designed to induce lower electricity use at times of high wholesale market prices or when system reliability is jeopardized" [3]. The most significant technological change on the demand side is the spread of smart meters and advanced metering infrastructure. According to this environment, traditional DSM programs should redesign to an automated market-based mechanism. The responses from the demand-side resources should also be reliable, fast, flexible, and large enough to compete with the supply-side resources. DSM programs can be classified into load management (LM) types and energy efficiency (EE) types [4–8].

In order to deal with peak load conditions, electric utilities have to invest in system capacity, which is underutilized during most times. Not surprisingly, utilities have been seeking methods to improve capacity utilization. Demand response (DR) is one mechanism utilities use to curtail or shift peak customer load [9–11].

Regarding the remuneration, a price-based and an incentive-based DR program are used. In a price-based DR program, consumers reduce their power consumption by responding to the electricity tariff set by the electricity market. In an incentive-based DR program, consumers are contracted by individuals or groups to reduce their power consumption for a certain period that the economic transactions requested in the electricity market.

For a price-based DR program, researchers investigated a multi-agent modeling and optimization algorithms under DR programs for real-time prices [12]; a coordination strategy between a micro network and a price-based demand response program for adjusting loads [13]; a bidirectional communication smart meter design between the household smart meters and the distribution management system [14]; a dynamic price scheme for electricity in a smart network, by analyzing the behavior and the possible demand response of the consumer [15]; and a modified real-time price model that encourages customer choice in electricity rates and is based on the amount of risk customers are willing to take and a real-time grid condition index developed by the California Independent Service Operator [16].

For an incentive-based DR program, research has shown that a fuzzy-based dynamic incentive scheme for residential customers can effectively incorporate the influences of socioeconomic conditions, expected curtailment, probability of curtailment, and notice period [17]. The dynamic adjustment of the offered prices is analyzed to reduce the demand and maximize its performance within T days [18]. The uncertainty of the demand in the network planning is modeled and includes the integral control of the load disconnection in search of the minimum cost [19]. The participation of smart homes with the help of a controller is capable of managing the electrical installation and restructuring the demand profile by changing the operation of the flexible loads [20]. The incentives motivate clients to participate in automatic DR programs with the purpose of compensating imbalances between offer and demand [21], and a DR program is shown from the economic point of view based on optimal incentives [22].

Demand resources have played an important role in Korea for more than 20 years. To reduce peak demand during summer and winter, DR programs and the operating system have been researched and implemented as a demonstration since 2010. In 2014, any customers who joined the DR market were able to sell their reduced demand in the electricity market as supply resources [23,24].

In this paper, a development and case studies of demand-side management solution (DSMS) in South Korea is presented. After that, the DSMS is verified with a one-year experience of the Korea DR program. To implement this solution, a design structure of the DSMS is proposed and tested. Customer baseline load (CBL), relative root mean squared error (RRMSE), load curtailment value, and money-saving of contracted customer's data are also displayed from the DSMS. To calculate CBL, a short historical period close to the event day was chosen, then, the CBL was calculated by merely averaging the data of the previous not-event days. After deciding the CBL, the assessment of the estimated CBL is needed. In order to verify the accuracy of the calculated CBL, RRMSE is used to assess the estimation error by comparing the actual electricity load and the estimated CBL. If the estimation error is close to zero, it means there is the high accuracy of the estimated CBL; if it is greater than zero, it means overestimated CBL; and if it is less than zero, it means underestimated CBL [25]. The calculation process of CB L and RRMSE are explained in Section 2.

This paper is composed as follows: Section 2 shows Materials (status of demand resource market in South Korea) and Methods (development of DSMS); Section 3 presents the money savings and energy conservation results using DSMS; finally, the discussion and conclusions are given in Sections 4 and 5.

2. Materials and Methods

2.1. Status of Demand Resources Market, South Korea

The role of the Korea Power Exchange (KPX) is to control the operation of South Korea's electricity market and the power system, as well as execute real-time dispatch and establish the basic plan for supply and demand. Every year, the KPX has issued an annual report for the electricity market trend and analysis. In this section, the status of the demand resource market from the 2016 annual report is summarized. Demand response refers to a suite of policies and institutions to provide efficient and stable electric power service at the lowest cost by helping consumers change their consumption pattern. Under the current contract-based utility rate schemes in South Korea, consumers have a very weak incentive to voluntarily participate in the demand response [26].

The introduction of the demand response program can effectively stabilize the electricity market and the operation of its system. The demand response can decrease investment in the power system including generation, transmission, and transformation networks; enhancing reliability in electric-power supply. Consumers can take part in the demand response programs by reducing their electricity usage at critical times through monitoring demand or securing a load that can be shut down by KPX and then making a bid on a load.

In early 2008, when the demand response market opened, it was bidding-based sponsored by the Electricity Industry Fund and, now, is making a transition into an advanced demand response market where market price and real-time system operation are linked with each other for resource transaction. In 2012, a smart demand response market opened where demand resources in small and medium quantity were traded in an effort to tap into the highly reliable and easily accessible demand resources using smart grid technology. The smart demand response market makes payment at a fixed rate on the condition of keeping the capacity to be curtailed unchanged. A payment is also made to those who cut down demand at the system operator's request for load curtailment. As the Electricity Business Act was revised to allow demand resource trading in the electricity market, in 2014, the demand response market and smart demand response market sponsored by the Electricity Industry Fund were abolished in late 2014, and the elimination of the two-month-ahead and week-ahead programs followed in late 2015 [26].

At the end of 2015, a new demand response market replaced these previous programs and was integrated into the electricity market. Figure 1 shows the process of the trading mechanism of a demand response market. The demand response market trades demand resources arranged by retailers, each demand resource is required to come from more than ten end users, and must be valued at above 10 MW. The DR aggregator collects consumers to organize demand resources. After registering with KPX, these resources are certified for trading under the same rules governing the centrally dispatched generators. Demand resources are put on a bid against power generation resources on a daily basis, and when sold, demand curtailment begins. In the system operation process, consumers are required to cut down on demand within an hour of a dispatch order.

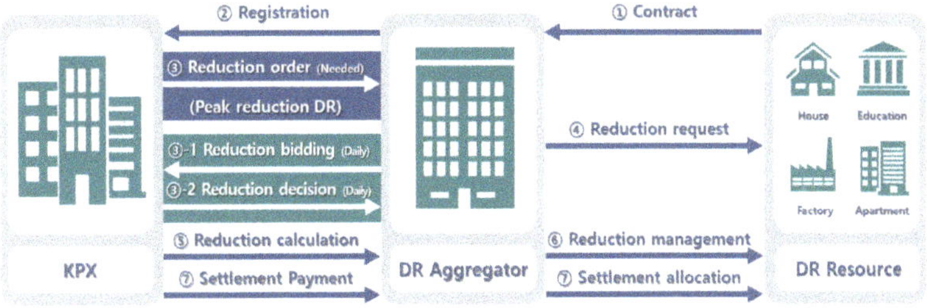

Figure 1. Trading mechanism of a demand response market.

The KPX calculates curtailed energy and makes payment to retailers who allocate the profits to consumers. As indicated in Tables 1 and 2, the demand response market has seen significant progress from both quantity and quality perspectives. The number of consumers has grown from 90 to 3592 and curtailed energy amounts to 175,771 MWh, up by 342.6 times from 513 MWh, becoming a world-renowned global demand response market. The transition to an advanced demand response market is also politically compelling. As the program matures along with demand resources integrated into the electricity market, market participants will feel easy to understand the system and the market will flourish. Furthermore, lower resistance towards generators and greenhouse gases are expected along with eased market exploitation and a more efficient market [23].

Table 1. Consumers and curtailed power before market opening [23].

Year	2010	2011	2012	2013	2014
No. of consumers	90	119	158	159	159
Capacity available for curtailment (MW)	2219	3049	3612	3615	3615
Curtailed energy (MWh)	513	690	785	556	682

Table 2. Consumers and curtailed power after market opening (provided by KPX, July 2018).

Year	2014.12~ 2015.05	2015.06~ 2015.11	2015.12~ 2016.05	2016.06~ 2016.11	2016.12~ 2017.05	2017.06~ 2017.11	2017.12~ 2018.05	2018.06~ 2018.11
No. of DR aggregators	11	15	14	15	14	17	20	22
No. of consumers	861	1323	1519	1970	2223	3195	3580	3592
Capacity available for Curtailment (MW)	1520	2444	2889	3272	3885	4352	4271	4222
Curtailed energy (MWh)	117,075	91,034	98,898	293,955	113,661	62,110	121,206	-

2.2. Development of Demand-Side Management Solution (DSMS)

To realize the trading mechanism of a demand response market, as shown in Figure 1, the structure of a demand-side management solution is designed, as shown in Figure 2. The DSMS directly captures usage data by sensor every 5 min from demand resources such as a house, building, apartment, and factory. After that, these usage data are sent to the KPX server every 5 min. If a power shortage occurs, the KPX sends a reduction order to the DSMS. When the request is generated from KPX, the DSMS calculates the CBL of the DR duration time and, then, contacts the DR resources to request the contracted power reduction amount, and after that, demand curtailment begins. In the system operation process, consumers are required to cut down on demand within an hour of the dispatch order. The DSMS also calculates the RRMSE (relative root mean square error), the CBL (customer baseline load) of the DR resources, regularly to keep the DR resources in the Korea electricity market.

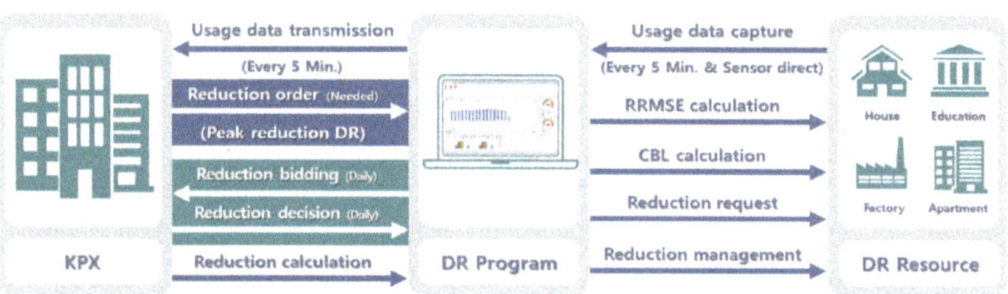

Figure 2. The structure of a demand-side management solution.

The CBL is the prediction of the amount of electricity that would generally be used if electricity consumption had not been reduced by the KPX directive [27]. Figure 3 shows the power consumption and CBL of the desired date and time. The bar graph illustrates power usage at 5 min, 10 min, and 30 min, respectively and the line means the CBL of every hour. At the bottom, the dialog box shows contracted curtail power, CBL, load, result of curtailed power, and remained contracted curtail power in order.

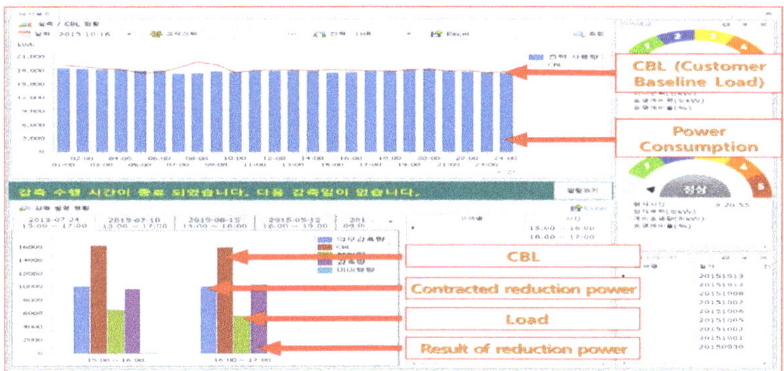

Figure 3. Dashboard of power usage.

Figure 4 illustrates a customer baseline load (CBL) calculation. The CBL is an average hourly energy consumption calculated as follows: According to the KPX guideline, the CBL calculation method is either MAX 4/5 or Mid 6/10. The Max 4/5 method is calculated using the electricity usage in normal working for 5 consecutive days. To calculate the CBL, first, the smallest electricity usage day of the 5 days is excluded, then, the average usage for 4 days is used as Max 4/5 CBL. The Mid 6/10 method is calculated based on the power electricity usage in normal working for 10 consecutive days. Two days are excluded from the top and bottom usage of the 10 days, respectively. The average usage of the remaining six days is used as Mid 6/10 CBL. Table 3 explains how to calculate the CBL as MAX 4/5. First, D-2 is the smallest electricity usage day, therefore, this day is excluded. Then, the average usages of the remained 4 days are added, and then divide by 4 [27].

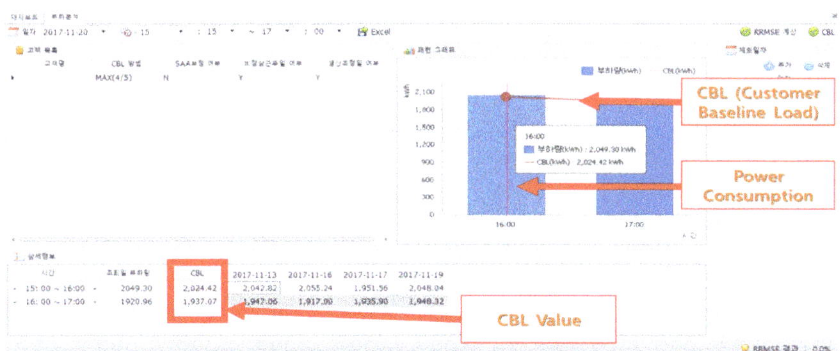

Figure 4. Customer baseline load calculation.

Table 3. Calculation example of the customer baseline load (CBL) (Max4/5 Method).

	Energy Consumed (kWh)	Selection
D-1	2048.04	O
D-2	1889.74	X
D-3	1951.56	O
D-4	2055.24	O
D-5	2042.82	O
CBL	Average of selection value = 2024.42	

To estimate the DR reduction value, the pattern of regular power use should be fairly uniform, and an objective evaluation technique is required. For this objective evaluation, the RRMSE was used as an index to specify the uniformity of the pattern of power usage. Figure 5 shows the RRMSE for the customer who wants to participate in demand response market and Equation (1) shows that RRMSE calculation [26]. In this equation, D is an investigation day, $D(n)$ is the number of investigation days, T is a time duration of an investigation day, $T(n)$ is number of time durations of an investigation day, $CBL_{d,t}$ is a customer baseline load at t hour on d day, and $Load_{d,t}$ is an electricity usage at t hour on d day.

$$\sqrt{\frac{\sum_{d \in D, t \in T}\left(CBL_{d,t} - Load_{d,t}\right)^2}{D(n) \times T(n)}} \div \frac{\sum_{d \in D, t \in T} Load_{d,t}}{D(n) \times T(n)} \tag{1}$$

$D(n)$: Number of investigation days
T: Time duration of investigation day
$T(n)$: Number of time durations of investigation day
$CBL_{d,t}$: Customer baseline load at t hour on d day
$Load_{d,t}$: Electricity usage at t hour on d day

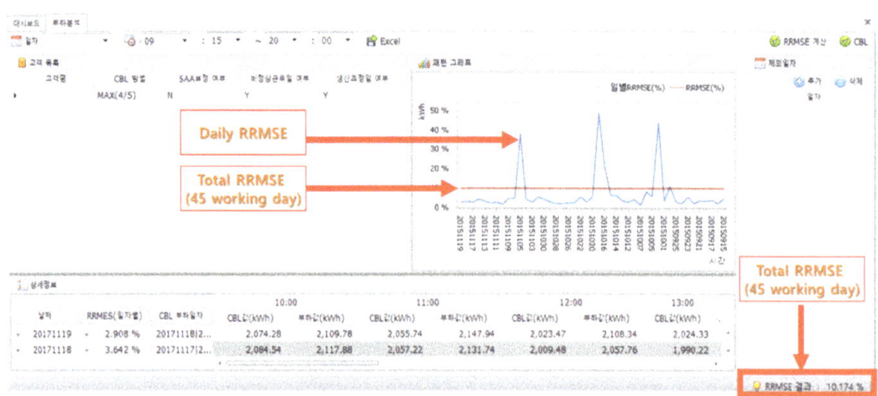

Figure 5. Relative root mean squared error (RRMSE) for a customer.

The RRMSE is calculated by dividing the RMSE (root mean square error) with the average value of electricity usage data. The fluctuation between the CBL and actual electricity usage is a critical judgment criterion as a reliable DR resource. To register as a DR resource in the Korea electricity market, the RRMSE value must be less than 30%; if the value exceeds 30%, it is not allowed to join the DR market. If the RRMSE value becomes more increased, conformity of power usage pattern decreases, which makes it difficult to judge the reduction value accurately. Although the U.S. PJM power DR market is set at less than 20%, the Korea DR market is set to within 30% at the beginning of the system. To determine the suitability of the DR customer, the KPX enforces an annual RRMSE assessment, and then that result determines whether the DR customer can participate in the DR market for one year. This calculation is based on data from 45 weekdays from the date of verification [26]. For example, the flat line of the electricity usage graph in Figure 5 indicates 10%, that means Figure 5 is a reliable DR resource, and therefore this customer can participate in the DR market.

The incentive for participating in the DR market can be classified into basic settlement and performance settlement, and the monthly basic settlement payment is as shown in Table 4. According to the reduction duration time, the actual performance-settlement payment is different under the KPX condition [27].

Table 4. Monthly basic settlement payment 2016–2018.

Month	Basic Settlement Payment [KRW/kW]	Weekday
2016.12	5186.22	22
2017.01	4994.68	20
2017.02	4475.45	20
2017.03	3767.84	22
2017.04	1462.24	20
2017.05	1335.86	20
2017.06	3858.63	21
2017.07	5395.95	21
2017.08	4930.56	22
2017.09	3694.44	21
2017.10	1213.57	17
2017.11	2919.16	22
2017.12	4,697.14	19
2018.01	5,763.12	22
2018.02	4,249.36	18
2018.03	3,519.93	21
2018.04	1,245.46	21
2018.05	1,036.46	21
2018.06	3,300.72	19
2018.07	5,803.64	22
2018.08	5,354.86	22
2018.09	3,150.83	17
2018.10	1,003.48	21
2018.11	2,717.80	22
Total	85,077,404	493

To calculate the basic settlement money, Equations in (2) are applied to the integrated demand-side management solution [26].

$$\mathrm{DRBP}_{i.m} = \mathrm{ORC}_{i.m} \times \mathrm{BP}_m \times 1,000$$
$$\mathrm{IBPC}_{i.m} = \frac{\mathrm{TDRBP}_i}{\mathrm{ORC}_{i.m} \times \mathrm{MRT}} \times \sum_{t}^{m} DRD_{i.t} \times 2 \times DF_{i,t}$$
$$DRD_{i.t} = Max(RSO_{i,t} \times 0.97 - DR_{i,t}, \, 0)$$
$$\mathrm{BPC}_{i.m} = Min(\mathrm{DRBP}_{i.m}, \, \mathrm{IBPC}_{i.m})$$
$$\mathrm{FDRBP}_{i.m} = \mathrm{DRBP}_{i.m} - \mathrm{BPC}_{i.m}$$

$$(2)$$

$DRBP_{i.m}$ Demand response basic payment by monthly (KRW)
$ORC_{i.m}$ Obligation reduction capacity (MW)
BP_m Basic price by monthly (KRW/kW)
$IBPC_{i.m}$ Initial basic penalty charge (KRW)
$TDRBP_i$ Total basic settlement money during the contract period (KRW)
MRT Maximum reduction time (Max 60 h)
$DRD_{i.t}$ Dispatch reduction deficiency (kWh)
$RSO_{i,t}$ Reduction ordered by system operator (MWh)
$DR_{i,t}$ Dispatched reduction (kWh)
$DF_{i,t}$ Dispatch flag (1 for active, 0 for non-active)
$BPC_{i.m}$ Basic penalty charge by monthly (KRW)
$FDRBP_{i.m}$ Final demand response basic payment by monthly (KRW)

To verify the DSMS functional test, the sampled data is divided into the following three categories: (1) large amount, (2) medium amount, and (3) small amount. Table 5 shows sampled data.

Table 5. The data for functional verification of demand-side management solution (DSMS).

Amount	Large	Medium	Small
Name	N Company	Provincial Government	G store
Type	Factory	Building	Retail
Peak (kW)	24,561	2002	57
Average power consumption (kW)	22,306	1534	52
Contracted capacity (kW)	10,000	1000	40

Baseline load, peak, average power consumption, and CBL are calculated based on the customer's power usage which is monitored and recorded from an electricity smart meter. Figures 6–8 show the data on 26 June 2017. The large amount, N company, is a chemical company located in the southwestern area of South Korea. The power consumption pattern of N company is a typical factory type. Figure 6 illustrates the power consumption pattern baseline load and peak. The medium amount, a provincial government building, is located in the central area of South Korea. The CBL of this customer is 1938.56 kW at 13:00~14:00 on 26 June 2017, as shown in Figure 7. Lastly, the small amount, G store, is a retail store located in the southeastern area of South Korea. Figure 8 shows the RRMSE value as 9.815%, less than 30%.

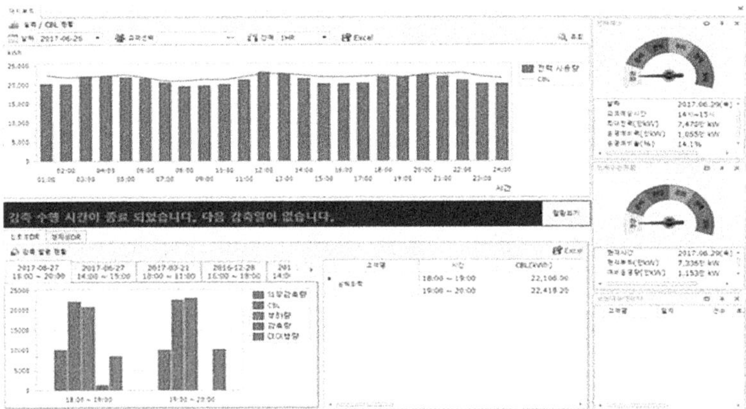

Figure 6. N company power consumption and baseline load.

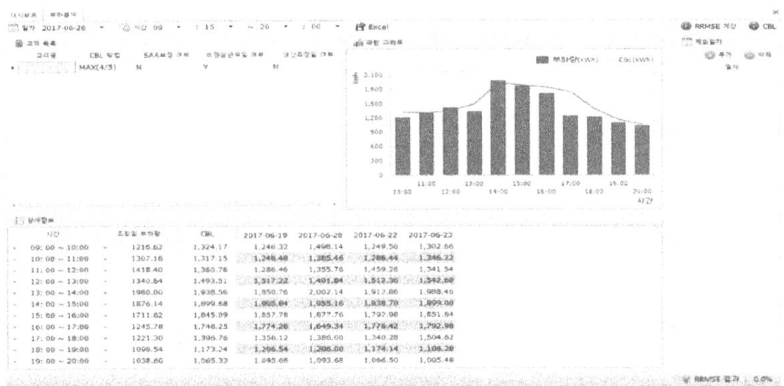

Figure 7. Provincial government building CBL calculation.

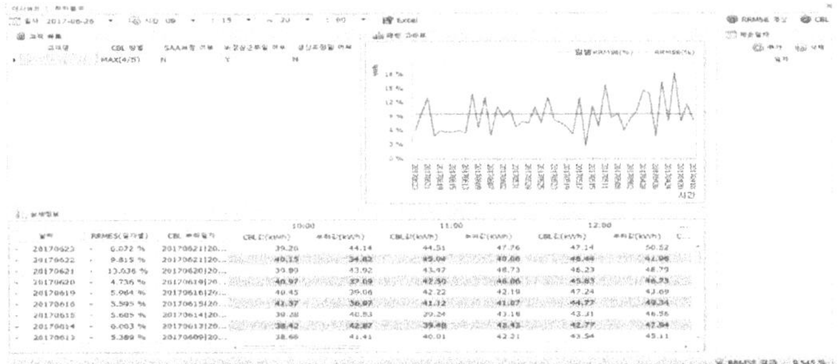

Figure 8. G store power RRMSE.

3. Results

Table 6 shows the DR benefit and reduction rate results for three contracted customers after participating in the DR program. The results show the money-saving and energy conservation using the DSMS. The data used in this paper was collected in two years (from December 2016 to November 2018) from selected customers in South Korea. The N company case is selected to demonstrate the result. In the case of N company, the contracted capacity is 10 MW, participating in the 20 h DR dispatch during the two years. As a result, the DR delivery rate which is an average 108.03% and the benefit from the DR participation which is 854,900,394 KRW (basic settlement benefit is 844,014,870 KRW and actual settlement money is 10,885,524 KRW) are occurring, and the savings in electricity cost over the two years is about 2,160,560 KRW

Table 6. Participation result of annual demand response market.

Name				N Company	
Month	Reduction Order	Reduction Duration Time	Basic-Settlement Benefit [KRW]	Performance-Settlement Benefit [KRW]	DR Delivery Rate [%]
2016.12	O	2 h	51,862,200	2,224,080	121.65%
2017.01	X	-	49,946,800	-	-
2017.02	X	-	44,754,500	-	-
2017.03	O	1 h	37,678,400	1,129,080	121.44%
2017.04	X	-	14,622,400	-	-
2017.05	X	-	13,358,600	-	-
2017.06	X	-	38,586,300	-	-
2017.07	O	3 h	53,221,870	2,429,646	95.30%
2017.08	X	-	49,305,600	-	-
2017.09	X	-	36,944,400	-	-
2017.10	X	-	11,421,800	-	-
2017.11	O	1 h	23,884,000	878,970	105.14%
2017.12	X	-	46,971,400	-	-
2018.01	O	9 h	57,631,200	1,056,280	108.17%
2018.02	O	1 h	42,493,600	1,094,716	114.69%
2018.03	X	-	35,199,300	-	-
2018.04	X	-	12,454,600	-	-
2018.05	X	-	10,364,600	-	-
2018.06	O	2 h	33,007,200	993,782	106.06%
2018.07	X	-	58,036,400	-	-
2018.08	X	-	53,548,600	-	-
2018.09	O	1 h	31,508,300	1,078,970	104.44%
2018.10	X	-	10,034,800	-	-
2018.11	X	-	27,178,000	-	-
	Total		844,014,870	10,885,524	-
Name				Provincial Government	
2016.12	O	2 h	5,186,220	211,700	114.32%
2017.01	X	-	4,994,680	-	-
2017.02	X	-	4,475,450	-	-
2017.03	O	1 h	3,767,840	99,475	105.72%
2017.04	X	-	1,462,240	-	-
2017.05	X	-	1,335,860	-	-

Table 6. *Cont.*

Name			Provincial Government		
2017.06	X	-	3,858,630	-	-
2017.07	O	3 h	5,395,950	336,334	127.28%
2017.08	X	-	4,930,560	-	-
2017.09	X	-	3,694,440	-	-
2017.10	X	-	1,142,180	-	-
2017.11	O	1 h	2,305,028	76,229	91.24%
2017.12	X	-	4,697,140	-	-
2018.01	O	9 h	5,763,120	119,026	121.89%
2018.02	O	1 h	4,249,360	131,721	138.00%
2018.03	X	-	3,519,930	-	-
2018.04	X	-	1,245,460	-	-
2018.05	X	-	1,036,460	-	-
2018.06	O	2 h	3,300,720	106,706	113.88%
2018.07	X	-	5,803,640	-	-
2018.08	X	-	5,354,860	-	-
2018.09	O	1 h	3,150,830	120,557	116.69%
2018.10	X	-	1,003,480	-	-
2018.11	X	-	2,717,800	-	-
	Total		84,391,878	1,201,747	-
Name			G Store		
2016.12	O	2 h	610,614	21,309	95.73%
2017.01	X	-	599,362	-	-
2017.02	X	-	537,054	-	-
2017.03	O	1 h	446,284	10,547	93.41%
2017.04	X	-	175,469	-	-
2017.05	X	-	160,303	-	-
2017.06	X	-	463,036	-	-
2017.07	O	3 h	647,514	36,711	170.65%
2017.08	X	-	591,667	-	-
2017.09	X	-	443,333	-	-
2017.10	X	-	137,062	-	-
2017.11	O	1 h	262,012	8,323	82.96%
2017.12	X	-	610,614	-	-
2018.01	O	9 h	599,362	11,935	122.22%
2018.02	O	1 h	537,054	10,158	106.42%
2018.03	X	-	446,284	-	-
2018.04	X	-	175,469	-	-
2018.05	X	-	160,303	-	-
2018.06	O	2 h	463,036	10,515	112.22%
2018.07	X	-	647,514	-	-
2018.08	X	-	591,667	-	-
2018.09	O	1 h	443,333	10,113	97.89%
2018.10	X	-	137,062	-	-
2018.11	X	-	262,012	-	-
	Total		10,147,42	119,611	-

To verify the capacity of the demand response from the customer, the request from KPX on 20 July 2017, 14:00~17:00, is displayed in Table 7. The large customer, N company, a big factory which has a contracted capacity of 10,000 kW delivered 97%, 93%, and 96% for each period and the average delivery rate was 95%. Figure 9 Illustrates the CBL load, the DR reduction result, and the delivery rate of each period. For example, Figure 9a shows the CBL is 18,059 kW, the real load is 8378 kW, the DR reduction is 9,681 kW, and the contracted capacity is 10,000 kW at 14:00~15:00.

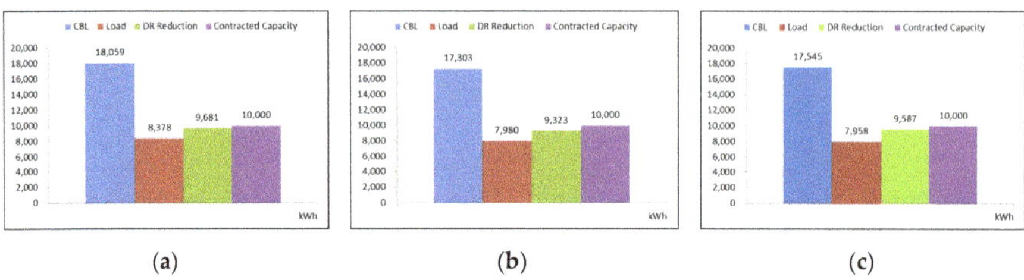

|(a)|(b)|(c)|

Figure 9. N company demand response result (data of 20 July 2017, 14:00~17:00).

Table 7. Result of demand response from customer.

Amount		Large	Medium	Small
Name		N Company	Provincial Government	G Store
Type		Factory	Building	Retail
Contracted capacity (kW)		10,000	1000	120
CBL (kW)	14:00~15:00	18,059	1944	356
	15:00~16:00	17,303	1904	349
	16:00~17:00	17,545	1882	302
Real load (kW)	14:00~15:00	8378	669	158
	15:00~16:00	7980	634	121
	16:00~17:00	7958	470	113
DR reduction result (kW)	14:00~15:00	9681	1275	198
	15:00~16:00	9323	1270	227
	16:00~17:00	9587	1412	189
DR delivery rate (%)	14:00~15:00	97%	128%	165%
	15:00~16:00	93%	127%	189%
	16:00~17:00	96%	141%	158%

4. Discussion

Unlike the traditional energy management models that focus on the supply side, the DSMS considers the energy demand and control on the interactions between customers and supplier to manage electricity usage reduction and money saving. Through the implementation of the DSMS technology, the end user can automatically check necessary information, such as the CBL, the RRMSE value, and the amount of DR reduction, without the need for complex formulas and contents. Furthermore, the CBL and current usage can be checked in real time by monitoring power usage every 5 min. The DSMS operated in the actual South Korea DR market for a year and based on these results the developed solution was verified. In addition, the result illustrates that the integrated demand-side management solution contributes by participating in the DR market and provides a benefit and satisfaction to the consumer.

The researcher and stakeholders of the DR market should consider the criteria value of the RRMSE. As previously mentioned in Section 2, the U.S. PJM power suggests a RRMSE value less than 20%, but the South Korea DR market sets the value within 30%. This consideration helps an effective DR operation and derives a successful outcome. This unique demand-side manage experience in South Korea could provide the rest of the world with a model to efficiently maintain a national power grid and potentially suggest the development of novel energy managing plans for local situations and policy.

5. Conclusions

In this paper, a case study of the demand-side management solution in South Korea is introduced and explained. The experience from the consumer and the DR aggregator shows that the integrated demand response solution technologies is a fast-responding approach in a cost-effective way. The curtailed energy from contracted customers contributed by reducing peak power in the national power grid and therefore can effectively provide a reliable power system. The demand resource can be an alternative to the redundant generation in short term such as 5 min. The case of a 10 MW contracted customer shows average 108.03% delivery rate and the total benefit of 854,900,394 KRW for two years. It also shows that all customers regardless of the amount of participation have responded well to 20-h DR dispatch during the two years. The results illustrate that the integrated demand-side management solution contributes by participating in the DR market and gives a benefit and satisfaction to the consumer.

Author Contributions: Methodology, W.K. and H.V.; software, H.-J.C. and S.-H.S.; validation, W.K. and H.-J.C.; formal analysis, S.-H.S.; investigation, W.K. and H.V.; resources, W.K., and H.-J.C.; data curation, W.K., and H.-J.C.; writing—original draft preparation, W.K., and H.-J.C.; writing—review and editing, W.K., and H.-J.C.; visualization, H.-J.C.; supervision, W.K. All authors have read and agreed to the published version of the manuscript.

Funding: The authors extend their appreciation to the Deanship of Scientific Research at King Saud University for funding this work through research group no. (RG-1439-028).

References

1. Rahimi, F.; Ipakchi, A. Demand Response as a Market Resource under the Smart Grid Paradigm. *IEEE Trans. Smart Grid* **2010**, *1*, 82–88. [CrossRef]
2. Vos, A. Effective business models for demand response under the smart grid paradigm. In Proceedings of the IEEE PES Power Systems Conference and Exposition PES 09, Seattle, WA, USA, 15–18 March 2009; Volume 1, p. 1.
3. USDOE. Benefits of Demand Response in Electricity Markets and Recommendations for Achieving Them. In *Report to the United State Congress Pursuant to the Section 1252 of the Energy Policy Act of 2005*; United State Department of Energy (USDOE): Washington, DC, USA, 2006.
4. Kim, J.; Nam, Y.; Hahn, T.; Hong, J. Demand Response Program Implementation Practices in Korea. In Proceedings of the 18th IFAC World Congress, Milano, Italy, 28 August–2 September 2011; pp. 3704–3707.
5. Lee, G.; Hong, W. A Study on the Adopting Decentralized Energy Supplying System in Urban Area. *Archit. Inst. Korea* **2007**, *23*, 239–246.
6. Rhee, C.; Park, J. Demand Resource Policy and Program Design for Electricity Market in Korea. In Proceedings of the ACEEE Summer Study on Energy Efficiency in Industry, Buffalo, NY, USA, 4–6 August 2015.
7. Won, J.-R.; Song, K.-B. An Analysis on Power Demand Reduction Effects of Demand Response Systems in the Smart Grid Environment in Korea. *J. Electr. Eng. Technol.* **2013**, *8*, 1296–1304. [CrossRef]
8. IEEE Guide for the Benefit Evaluation of Electric Power Grid Customer Demand Response. In *IEEE Standards 2030.6-2016*; IEEE: New York, NY, USA, 2016; pp. 1–42.
9. Zhang, Y.; Chen, W.; Xu, R.; Black, J. A Cluster-Based Method for Calculating Baselines for Residential Loads. *IEEE Trans. Smart Grid* **2016**, *7*, 2368–2377. [CrossRef]
10. Albadi, M.H.; El-Saadany, E.F. Demand Response in electricity markets: An Overview. In Proceedings of the IEEE Power Engineering Society General Meeting, Tampa, FL, USA, 24–28 June 2007; pp. 1–5.

11. Conejo, A.J.; Morales, J.M. Real-time demand response model. *IEEE Trans. Smart Grid* **2010**, *1*, 236–242. [CrossRef]
12. Wang, Z.; Raman, P. An Evaluation of Electric Vehicle Penetration under Demand Response in a Multi-Agent Based Simulation. In Proceedings of the 2014 IEEE Electrical Power and Energy Conference (EPEC), Calgary, AB, Canada, 12–14 November 2014.
13. Zhang, C.; Xu, Y.; Dong, Z.; Wong, K.P. Robust Coordination of Distributed Generation and Price—Based Demand Response in Microgrids. *IEEE Trans. Smart Grid* **2017**, *9*, 4236–4247. [CrossRef]
14. Minchala-avila, L.I.; Armijos, J.; Pesántez, D.; Zhang, Y. Design and Implementation of a Smart Meter with Demand Response Capabilities. *Energy Procedia* **2016**, *103*, 195–200. [CrossRef]
15. Sachdev, R.S.; Singh, O. Consumer's Demand Response to Dynamic Pricing of Electricity in a Smart Grid. In Proceedings of the 2016 International Conference on Control, Computing, Communication and Materials (ICCCCM), Allahbad, India, 21–22 October 2016.
16. Vanderkley, T.S.; Negash, A.I.; Kirschen, D.S. Analysis of Dynamic Retail Electricity Rates and Domestic Demand Response Programs. In Proceedings of the 2014 IEEE Conference on Technologies for Sustainability (SusTech), Portland, OR, USA, 24–26 July 2014; pp. 172–177.
17. Gadham, K.R.; Ghose, T. Design of Incentive Price for Voluntary Demand Response Programs Using Fuzzy System. In Proceedings of the International Conference on Electrical Power and Energy Systems (ICEPES), Bhopal, India, 14–16 December 2016; pp. 363–367.
18. Khezeli, K.; Bitar, E. Risk-Sensitive Learning and Pricing for Demand Response. *IEEE Trans. Smart Grid* **2017**, *9*, 6000–6007. [CrossRef]
19. Schachter, J.A.; Mancarella, P.; Moriarty, J.; Shaw, R. Flexible Investment under Uncertainty in Smart Distribution Networks with Demand Side Response: Assessment Framework and Practical Implementation. *Energy Policy* **2016**, *97*, 439–449. [CrossRef]
20. Sarris, T.; Messini, G.; Hatziargyriou, N. Residential demand response with low cost smart load controllers Mediterr. In Proceedings of the Mediterranean Conference on Power Generation, Transmission, Distribution and Energy Conversion (MedPower 2016), Belgrade, Serbia, 6–9 November 2016.
21. Jain, M.; Chandan, V.; Minou, M.; Thanos, G.; Wijaya, T.K.; Lindt, A.; Gylling, A. Methodologies for Effective Demand Response Messaging. In Proceedings of the 2015 IEEE International Conference on Smart Grid Communications (SmartGridComm), Miami, FL, USA, 2–5 November 2015; pp. 453–458.
22. Wijaya, T.K.; Vasirani, M.; Villumsen, J.C.; Aberer, K. An Economic Analysis of Pervasive, Incentive-Based Demand Response. In Proceedings of the 2015 IEEE International Conference on Smart Grid Communications (SmartGridComm), Miami, FL, USA, 2–5 November 2015; pp. 331–337.
23. "Electricity Market Trends & Analysis", Annual Report, KPX. Available online: http://www.kpx.or.kr/eng/downloadBbsFile.do?atchmnflNo=23219 (accessed on 1 December 2018).
24. Lee, S.S.; Lee, H.C.; Yoo, T.H.; Noh, J.W.; Na, Y.J.; Park, J.K.; Yoon, Y.T. Demand Response Prospects in the South Korean Power System. In Proceedings of the IEEE PES General Meeting, Providence, RI, USA, 25–29 July 2010.
25. LEEa, Seungman. "COMPARING METHODS FOR CUSTOMER BASELINE LOAD ESTIMATION FOR RESIDENTIAL DEMAND RESPONSE IN SOUTH KOREA AND FRANCE: PREDICTIVE POWER AND POLICY IMPLICATIONS". 2019. Available online: https://www.researchgate.net/publication/337948056_Comparing_Methods_for_Customer_Baseline_Load_Estimation_for_Residential_Demand_Response_in_South_Korea_and_France_Predictive_Power_and_Policy_Implications (accessed on 1 March 2020).
26. *Intelligent Demand Resource Market Operating Rules and Guideline*; Korea Power Exchange: Jeollanam-do, Korea, 2016. (Printed in Korean)
27. Lee, J.; Yoo, S.; Kim, J.; Song, D.; Jeong, H. Improvements to the customer baseline load (CBL) using standard energy consumption considering energy efficiency and demand response. *Energy* **2018**, *144*, 1052–1063. [CrossRef]

 *applied sciences*

Article

Optimal Operational Scheduling of Distribution Network with Microgrid via Bi-Level Optimization Model with Energy Band

Ho-Young Kim [1], Mun-Kyeom Kim [2,*] and Hyung-Joon Kim [2]

[1] Basic Research Center for Electric Power, KEPCO Research Institute, Bldg 130, Seoul National University, 1 Gwanak-ro, Gwanak-gu, Seoul 08826, Korea; duaud32@naver.com
[2] Department of Energy System Engineering, Chung-Ang University, 84 Heukseok-ro, Dongjak-gu, Seoul 156-756, Korea; glen625@cau.ac.kr
* Correspondence: mkim@cau.ac.kr; Tel.: +82-2-820-5271

Received: 19 August 2019; Accepted: 3 October 2019; Published: 10 October 2019

Abstract: An optimal operation of new distributed energy resources can significantly advance the performance of power systems, including distribution network (DN). However, increased penetration of renewable energy may negatively affect the system performance under certain conditions. From a system operator perspective, the tie-line control strategy may aid in overcoming various problems regarding increased renewable penetration. We propose a bi-level optimization model incorporating an energy band operation scheme to ensure cooperation between DN and microgrid (MG). The bi-level formulation for the cooperation problem consists of the cost minimization of the DN and profit maximization of the MG. The goal of the upper-level is to minimize the operating costs of the DN by accounting for feedback information, including the operating costs of the MG and energy band. The lower-level aims to maximize the MG profit, simultaneously satisfying the reliability and economic targets imposed in the scheduling requirements by the DN system operator. The bi-level optimization model is solved using an advanced method based on the modified non-dominated sorting genetic algorithm II. Based on simulation results using a typical MG and an actual power system, we demonstrate the applicability, effectiveness, and validity of the proposed bi-level optimization model.

Keywords: bi-level optimization model; cooperation; distributed energy resource; distribution network; energy band; microgrid; modified non-dominated sorting genetic algorithm II

---

1. Introduction

1.1. Background

In recent years, significant global efforts have been devoted to developing renewable energy sources to reduce the demand for fossil fuels as well as to limit carbon emissions and air pollution. In particular, distributed and small-scale wind and solar photovoltaic (PV) power generation systems have undergone dramatic growth [1,2]. Furthermore, microgrids (MGs) have attracted attention owing to their potential to provide electricity in a reliable, economical, efficient, and environmentally-friendly manner from distributed energy resources (DERs) [3,4]. MGs provide an effective means of overcoming the intermittency of DERs and enabling bidirectional transactions. Moreover, they contribute to limiting carbon emissions, allow for the diversification of energy sources, and reduce cost [5].

At present, the majority of electrical energy consumed is provided by nuclear or fuel power plants with high capacities and reliability. However, the costs of investing in such power plants remains very high. Hence, DERs implemented at a smaller scale have been attracting considerable interest, as they incur smaller capital requirements as well as have a lower environmental impact [6]. However, the increased use of renewable energy and high-efficiency distributed generation (DG) sources in power systems has resulted in power generation systems becoming smaller. Furthermore, these power systems are located closer to consumers on MGs, which comprise distribution networks (DNs) for a range of energy resources such as fuel cells, wind turbines (WTs), combined heat and power (CHP) systems, and PV systems [7]. The incorporation of DG sources into a network offers numerous benefits with respect to the overall performance, provided that these sources are optimally scheduled and coordinated [8]. An important challenge regarding MGs is the optimization of their operation, which involves determining the best means of generating value for each unit to maximize the profit. In this regard, DG sources that are capable of CHP production are being used extensively in MGs as well as with DERs that are stochastic in nature [9]. Since power generation by DERs includes an element of uncertainty, it is necessary to forecast their output. This topic formed part of a previous study in which the load uncertainties and those associated with DERs were considered [10].

1.2. Literature Review and Motivation

Studies in this field should take into account the effects of the operational performance in MGs. However, numerous researches on MGs have only focused on the optimization of their operations. In such studies, the optimization of the operational cost as well energy savings and emission reductions are the primary objectives, and different algorithms are employed to solve the optimization problem. A method for the economic dispatch optimization of MGs, which considers the load as well as microturbine (MT) constraints was proposed with the aim of minimizing the fuel cost [11]. Low-cost operations and energy management are generally considered as necessary for MGs to ensure that they can meet the power demand, facilitate improved penetration levels of renewable energy, and allow for control over the power exchanged with the utility grid [12,13]. The power flow and cost management of MGs have been studied in depth [14,15]. A decentralized architecture for multiagent systems has also been proposed for the economic dispatch of MG [16]. However, their operation is complicated by the bidirectional energy exchange between the MGs and DN [17]. Therefore, further research on power system operation is necessary to determine a suitable approach for cooperation between the DN and MG.

Furthermore, various uncertainties pertaining to the economic operation of MGs have been considered. These factors are based on the assumption that low-voltage DNs sell energy to the MGs at real-time pricing tariffs [18]. In addition to these studies, which have only focused on the economic operation of MGs, several works have explored the benefits of using MGs with the DN. In this regard, an integrated solution that considered both the load dispatch of the MGs and reconfiguration of the main grid was proposed in an attempt to minimize the total operational costs of the main grid with multiple MGs [19]. Moreover, a co-optimization planning model for MGs was proposed, which considered the reliability of the power system as well as several economic criteria relating to the generation and transmission systems and the MG [20]. However, these co-optimization problems cannot be resolved directly using conventional optimal algorithms. Optimal operation of the entire power system should enable the decision makers to optimize their respective objective functions independently while simultaneously cooperating with one another. Thus, several researchers have studied bi-level optimization models to address this issue [21–23]. However, only the DN operation has been optimized in these studies, without the tie-line control being considered in the analysis. Therefore, further research relating to the decision-making framework that considers the tie-line control is required to achieve optimal cooperation between the DN and MG.

Recent work has also introduced various structures and methods for optimizing the operation of MGs, including approaches that use various optimization algorithms for MGs with different DERs. In particular, attempts at optimizing MG operations using a mixed integer nonlinear programming (MINLP) model have been reported, in which the aim was to minimize an objective function that considered the initial investment, operations, repair and maintenance, and environmental costs [24].

However, the mathematical solutions for models, such as that based on MINLP cannot be used to optimize large-scale nonlinear problems, which must be addressed using heuristic techniques. As a solution to this problem, the non-dominated sorting genetic algorithm II (NSGA-II) was used to allocate power to the units in the power system economically [25]. Although this algorithm incorporates several advanced concepts, including elitism, fast non-dominated sorting, and diversity maintenance along the Pareto solution, it still exhibits shortcomings in sustaining lateral diversity and acquiring the Pareto solution with high uniformity.

1.3. Contribution and Organization of Paper

This paper presents a bi-level optimization model to determine the optimal operation strategy for both the DN and MG. The upper-level optimization determines the scheduling requirement injected from the DN to the MG by minimizing the operating costs of the DN. The lower-level optimization provides feedback information regarding the received scheduling requirement by maximizing the MG profit while considering the energy band operational scheme. Moreover, a modified version of the NSGA-II (MNSGA-II) is applied to solve the model, resulting in improvements in the profit of the MG while reducing the DN operational cost. Simulations are performed on an IEEE (Institute of Electrical and Electronics Engineers) test system and actual system to validate the model and highlight its advantages over other multi-objective approaches.

The contributions of this study are as follows:

- Strategic behavior is proposed for cooperation between each operator for both the DN and MG based on the tie-line control strategy using a power margin (energy band operation scheme) of the tie-line between the DN and MG, which aids in providing a reliable and economic reference for responsibility sharing between two operators.
- A bi-level optimization model with an operation scheme based on the energy band is presented for cooperation between the DN and MG, in which the upper-level sends/receives the scheduling requirement/feedback information to/from the lower-level to determine the amount of energy of the tie-line.
- Finally, the MNSGA-II is employed, which preserves the diversity of the non-dominated solution laterally, and yields a Pareto solution with high uniformity, owing to the trade-off between the operational cost of the DN and the profit of the MG resulting in improved performance, as well as faster convergence and divergence.

The remainder of this paper is organized as follows: Section 2 addresses the methodology for the energy band. The bi-level optimization framework and multi-objective formulation is discussed in Section 3. Section 4 describes the proposed solution scheme and optimization process for cooperation based on the MNSGA-II. Section 5 presents a discussion of the obtained results. Finally, the paper is concluded in Section 6.

2. Methodology

2.1. Energy Band Operation Scheme

In general, the load exhibits relatively sudden fluctuations owing to the presence of renewable customers with a massive load in the grid-connected MG. The operational efficiency can be improved, and optimization can be achieved by sharing the operational information with the DER and the demand between the DN and MG system operator (DSO and MGO). The responsibility for balancing the supply–demand problem is transferred to the DSO when the MG is grid-connected. We used an operational scheme based on an energy band to divide the responsibility for balancing the supply and demand. This concept is based on a modification of the operational scheme originating from the frequency control band in the reserve capacity market [26]. An energy band operation scheme of a tie-line flow is illustrated in Figure 1. Here, the contractual tie-line flow refers to the existing contract power between the DN and the MG. The energy band is the marginal power of the existing contractual tie-line flow. On the other hand, the rescheduled tie-line flow indicates the modified tie line flow through the energy band from the contractual tie-line flow. The DSO and MGO can regulate the cost for the tie-line flow so that it remains reasonable, ensuring stable operation of both grids.

The rescheduled tie-line flow, P^*_{tie}, is expressed as:

$$P^*_{tie,c} - P^{EB}_{tie} \leq P^*_{tie} \leq P^*_{tie,c} + P^{EB}_{tie} \tag{1}$$

$$P^*_{tie,c}(t+1) - P^{EB}_{tie}(t) - P^*_{tie}(t) \leq u^*_{tie}(t) \leq P^*_{tie,c}(t+1) + P^{EB}_{tie}(t) - P^*_{tie}(t) \tag{2}$$

where P^*_{tie} is the rescheduled tie-line flow; $P_{tie,c}$ is the contractual tie-line power flow between the DN and MG; P^{EB}_{tie} is the size of the energy band; u^*_{tie} is the control signal for the tie-line flow.

The energy band constraint (Equation (2)) is introduced under the bi-level optimization model and rescheduled tie-line flow given in (Equation (1)), to prevent sudden changes in the tie-line flow. This scheme does not provide for the additional energy cost for a breach of the contractual tie-line flow within the energy band. However, the rescheduling information should be identified by the MGO prior to changing the feedback information from $P_{tie.c}$, because it is a schedule for the DER of the MG. This information can assist the operators in sharing the responsibility and maintaining stable operation.

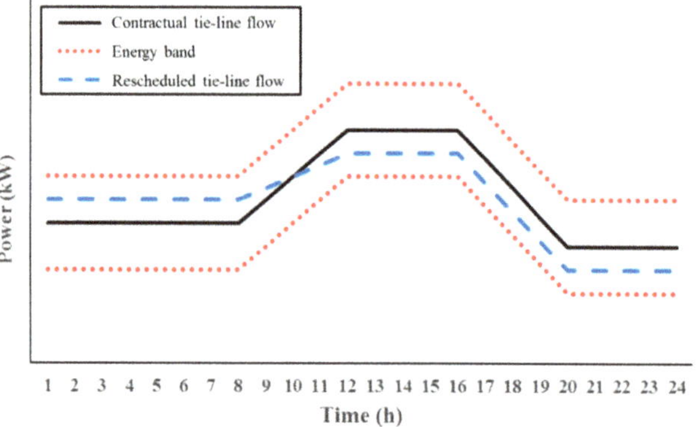

Figure 1. Energy band operational scheme.

2.2. Ramping Capability

At each time t, the ramping capability of upward and downward for the next period, $t + 1$ is calculated as follows:

$$r^*_u(t) = \left[\begin{array}{c} \sum\limits_{i=1}^{N_{Gm}} \min(P_{Gmi,max} - P^*_{Gmi}, u_{Gmi,max}) + \min(P^*_{tie,c}(t+1) + P^{EB}_{tie} - P^*_{tie}(t), P_{tie,max} - P^*_{tie}(t)) \\ + \sum\limits_{i=1}^{N_B} \min\left(\frac{1}{\eta_{di}}(-S_{Bi,min} + S^*_{Bi}(t)) - P_{Bi,min} + P^*_{Bi}(t)\right) \end{array} \right] \tag{3}$$

$$r^*_d(t) = \left[\begin{array}{c} \sum\limits_{i=1}^{N_{Gm}} \max(P_{Gmi,min} - P^*_{Gmi}(t), u_{Gmi,min}) + \max(P^*_{tie,c}(t+1) - P^{EB}_{tie} - P^*_{tie}(t), P_{tie,min} - P^*_{tie}(t)) \\ + \sum\limits_{i=1}^{N_B} \max\left(\frac{1}{\eta_{ci}}(-S_{Bi,max} + S^*_{Bi}(t)) - P_{Bi,max} + P^*_{B}(t)\right) \end{array} \right] \tag{4}$$

where $S_{Bi,t}$ is the state of charge in ith battery energy storage system (BESS) at time t.

$$\max(a,b) = \left\{ y \middle| y \geq a \text{ and } y \geq b, \ y \in \{a,b\} \right\} \tag{5}$$

$$\min(a,b) = \left\{ y \middle| y \leq a \text{ and } y \leq b, \ y \in \{a,b\} \right\} \tag{6}$$

By securing r^*_u, the MGO can mitigate the operational risk arising from energy shortages. Equation (4) corresponds to the dissipation of energy at time $t + 1$. Once the ramping capability has been determined, it is necessary to consider the contractual tie-line flow and energy band, because a change in the contractual tie-line flow will affect the MG demand patterns. The tie-line flow at time k can be set to the maximum value, $P^*_{tie,c} + P_{tie}{}^{EB}$, when the MG operating cost is higher than that of the DN. In this situation, the upward ramping capability from the tie-line flow may have a negative value when $P^*_{tie,c}$ $(t + 1)$ is lower than $P^*_{tie,c}$. Consequently, the values corresponding to the upward and downward ramping capability should reflect the changes in the contractual tie-line flow and energy band over the predictive horizons.

3. Problem Formulation

3.1. Bi-Level Optimization Model

In general, a bi-level optimization model is a decision model with a two-level structure and multiple participants [27]. The upper-level decision regulates the lower-level behavior while the optimal strategy of the lower-level influences the upper-level decision-making. Mathematically, the bi-level optimization model can be expressed as a pre-defined objective function subject to a set of static physical and operating limits, with its compact form being as follows:

$$
\begin{cases}
Upper-level: & \begin{cases} \underset{U}{Min}\ \omega_u F(Z) \\ s.t.\quad G(U,\ X) \le 0 \end{cases} \\
Lower-level: & \begin{cases} \underset{X}{Max}\ \omega_l f(Z) \\ s.t.\quad g(U,X) \le 0 \end{cases}
\end{cases}
\tag{7}
$$

where $F(Z)$ and $f(Z)$ are the objective functions of the upper and lower levels, respectively; $G(U,X)$ and $g(U,X)$ are the vector functions representing the equality constraints of the upper and lower levels, respectively; U and X are the decision variables of the upper and lower levels, respectively. ω_u and ω_l are weighting factors of upper-level and lower-level, respectively.

The bi-level decisions influence and constrain one another, because the optimization model can express the hierarchical relationship between the two levels. Figure 2 presents the overall scheme of the bi-level optimization model which is designed to incorporate the energy band during the problem formulation. The MG is integrated into the DN, and the DSO provides power to the MGO to balance the load. Furthermore, the WT, PV panel, MT, and battery energy storage system (BESS) are connected to different load nodes of the MG. In this case, the power flow between the DN and MG is bidirectional. The MG can not only purchase energy from the DSO but also sell energy to it. As shown in Figure 2, the DSO incentivizes the MGO to lower the cost of the energy it supplies using the optimal scheduling information for the tie-line exchange, and the MGO operates economically and safely by following the scheduling requirements. On the upper-level, the DSO who has the responsibility of operating the DN in an optimal fashion, optimizes the power exchanged between the DSO and MGO, so that its operating cost is minimized. In response, the lower-level determines the rescheduled tie-line flow with the energy band as feedback information. The exchanged power of the tie-line between the DN and MG is considered as coupling variables between the DSO and MGO, which determine the amount of the tie-line purchased power by the DSO at the upper-level in consideration of the tie-line exchange power for MGO's profit for stable operation of the DN, and it forces the decision maker to consider a multi-objective optimization problem. The DSO wants to buy less transaction levels from the MGO to minimize operating costs for optimal operation of the DN. Conversely, the MGO want to sell them to the DSO to maximize profits in the DN. These relationships are trade-off (i.e., minimized operating cost for DSO and maximized profit for MGO), and the solutions must be solved by a multi-objective problem in order to obtain a Pareto-front between the two operators. Thus, the proposed approach provides the optimal solution for cooperation between the DSO and MGO to help in decision-making, whenever there is a trade-off between operating cost of the DSO and profit of the MGO in the DN.

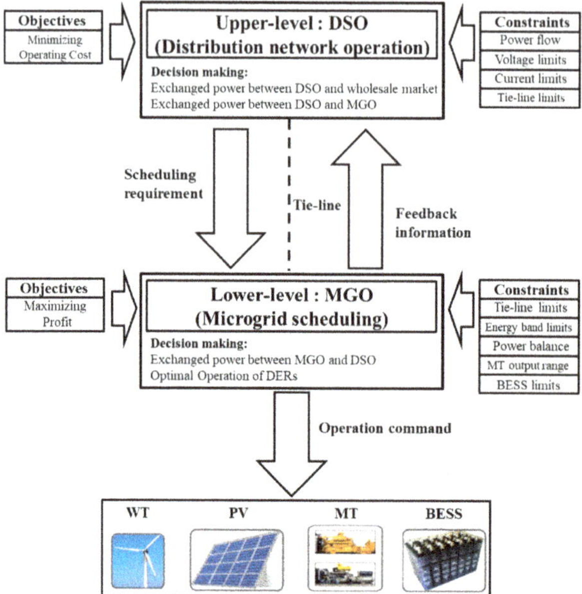

Figure 2. Overall scheme of a bi-level model. MGO: microgrid operator; DSO: distribution network operator; WT: wind turbine; PV: photovoltaic; MT: microturbine; BESS: battery energy storage system.

3.2. Upper-Level Model for DSO

To optimize the DN operations, the objective function of the upper-level model aims to minimize the operating cost from the DSO perspective, including three terms. The first term represents the power losses, while the second term indicates the cost of purchasing/selling active power from/to the day-ahead wholesale market, and the third term is the cost of exchanging power at the tie-line between the DN and MG. If $P_{tie} > 0$, the DSO is selling power to the MG and if $P_{tie} < 0$, the DSO is purchasing power from the MG, and also if $P_{tie} = 0$, no power exchange takes place between the DSO and MGO.

$$\min F = \sum_{t=1}^{T} [C_{loss}(t) + C_M(t) + C_{tie}(t)] \tag{8}$$

where $C_{loss}(t)$ is the cost of the power losses at time t; $C_M(t)$ is the cost of purchasing/selling active power from/to the day-ahead wholesale market at time t; $C_{tie}(t)$ is the cost of exchanging power at the tie-line between DN and MG at time t.

In general, the DN operating costs for renewable and storage energy are not considered when calculating the minimal operating cost, because the fuel costs are almost equal to zero. Therefore, these operating costs are not considered in our work. The detailed objective function is expressed as follows:

$$\min F = \sum_{t=1}^{T} \left[\rho_{loss} P_{loss}(t) + \rho_M(t)\left(P_{M,p}(t) - P_{M,s}(t)\right) + \rho_e(t) P_{tie}(t) \right] \tag{9}$$

where ρ_{loss} is the price for power losses, in \$/kWh; $P_{loss}(t)$ is the amount of active power losses; $\rho_M(t)$ is the day-ahead clearing price in the wholesale market at time t, in \$/kWh; $P_{M,p}(t)/P_{M,s}(t)$ is the power purchased from/sold to the wholesale market at time t, in \$/kWh; $\rho_e(t)$ is the day-ahead energy exchange price announced by the DSO for the MG at time t, in \$/kWh. Here, $P_M(t)$ and $\rho_e(t)$ are decision variables on the upper-level.

The power flow equations for the DN, including the active and reactive power, are modified as follow:

$$
\begin{cases}
P_{Gmi,t} - P_{Lmi,t} - V_{i,t} \sum_{j=1}^{n-1} V_{j,t}(G_{ij} \cos \theta_{ij,t} + B_{ij} \sin \theta_{ij,t}) = 0 \\
Q_{Gmi,t} - Q_{Lmi,t} - V_{i,t} \sum_{j=1}^{n-1} V_{j,t}(G_{ij} \sin \theta_{ij,t} - B_{ij} \cos \theta_{ij,t}) = 0
\end{cases}
\tag{10}
$$

where $P_{Gmi,t}$ is the active power of the ith node with the MG at time t; $P_{Lmi,t}$ is the active power of the ith load node at time t; $V_{i,t}$ is the voltage of the ith node at time t; $V_{j,t}$ is the voltage of the jth node at time t; G_{ij} is the conductance element of the DN admittance matrix; B_{ij} is the susceptance element of the DN admittance matrix; $\theta_{ij,t}$ is the phase angle difference between the ith and jth nodes at time t.

The amount of active power loss is calculated as follows:

$$
P_{loss}(t) = \sum_{i,j=1}^{N} R_{ij} \left(\frac{P_{Gmij,t}^2 + Q_{Gmij,t}^2}{V_{i,t}} \right)
\tag{11}
$$

where R_{ij} is the resistance of branch ij.

The DN power balance equation indicates that the sum of the power exchanged with the MG and power purchased from the market is equal to the sum of the demand and loss, as follows:

$$
\sum_{i=1}^{Nm} P_{tie,i}(t) + P_M(t) = P_d(t) + P_{loss}(t)
\tag{12}
$$

The inequality constraints represent the DN physical and security limits, and include the following:

- Voltage limits

$$
V_{i,min} \leq V_{i,t} \leq V_{i,max}, \quad \forall v_{slack} = 1
\tag{13}
$$

- Line current limits

$$
\frac{P_{Gmij,t}^2 + Q_{Gmij}^2}{V_{i,t}} \leq I_{max}^2
\tag{14}
$$

- Grid tie-line flow limits

$$
P_{tie,p,min} \leq P_{tie,p,t} \leq P_{tie,p,max}
\tag{15}
$$

$$
P_{tie,s,min} \leq P_{tie,s,t} \leq P_{tie,s,max}
\tag{16}
$$

- Price of power exchange limits

$$
0 \leq \rho_e(t) \leq \rho_{e,max}(t)
\tag{17}
$$

- Exchanged power limit with wholesale market

$$
P_M(t) \leq P_{M,max}(t)
\tag{18}
$$

3.3. Lower-Level Model for MG

The lower-level objective function is intended to maximize the profit of the MGO connected to the DN considering five different terms.

$$
\max f = \sum_{t=1}^{T} [Pr_L(t) + Pr_e(t) + Pr_m(t) + Pr_{PV}(t) + Pr_{BESS}(t)]
\tag{19}
$$

where $Pr_O(t)$ is the profit including loads, and exchanged power between DSO and MGO, MT, PV, and BESS, respectively.

The profit terms of the above objective function are defined as follows:

$$Pr_L(t) = \rho_e(t) \sum_{i=1}^{N} P_{Li}(t) \tag{20}$$

$$Pr_e(t) = \rho_e(t)\left(P_{tie,p}(t) - P_{tie,s}(t)\right) \tag{21}$$

$$Pr_m(t) = -\sum_{i}^{N} \rho_{mi} P_{Gmi}(t) \tag{22}$$

$$Pr_{PV}(t) = -\sum_{i}^{N} \rho_{PVi} P_{PV}(t) \tag{23}$$

$$Pr_{BESS}(t) = -\sum_{i}^{N} \rho_{BESSi}\left(P_{B,d}(t) - P_{B,c}(t)\right) \tag{24}$$

where $P_{B,c}$ and $P_{B,d}$ are the battery charge and discharge power, respectively.

The micro sources in the DN include the WTs, PV panels, and BESS. The models and equations for these micro sources are presented below:

(1) WTs

The output power of the WTs is modeled using the following parameters, provided in [27], which can be expressed as

$$P_{WT} = \begin{cases} 0 & v \leq v_{ci} \; or \; v \geq v_{co} \\ \varepsilon_1 \cdot v^3 - \varepsilon_2 \cdot P_{rate} & v_{ci} < v \leq v_r \\ P_{rate} & v_r < v \leq v_{co} \end{cases} \tag{25}$$

where v is wind speed; v_{ci} is the cut-in speed; v_{co} is the cut-off speed; v_r is the rated speed; ε_1 and ε_2 are the fitting parameters of the WT power curve; P_{rate} is the rated output power of the WT.

(2) PV panels

The PV output power is expressed as a function of the irradiance and temperature. Thus, the PV panels can be modeled as

$$P_{PV} = P_{STC,max} \cdot \frac{G_{AC}}{G_{STC}} \cdot (1 + k(T_e - T_{STC})) \tag{26}$$

where P_{PV} is the output power of the PV system; $P_{STC,max}$ is the maximum output under standard test conditions; G_{AC} is the current irradiance; G_{STC} is the standard irradiance; k is the temperature coefficient; T_e is the current temperature; T_{STC} is the standard temperature; G_{STC} = 1000 W/m^2, and T_{STC} = 25 °C [28].

(3) BESS

The energy storage units are used for energy compensation between the MG supply and demand. The following constraints are considered for the charging and discharging strategy [29]:

$$S_{Bi,t,min} \leq S_{Bi,t} \leq S_{Bi,t,max} \tag{27}$$

$$S_{Bi,t+1} = S_{Bi,t} + \eta_c P_{B,c}(t)\Delta t - P_{B,d}(t)\Delta t / \eta_d \tag{28}$$

where η_c and η_d are the battery efficiencies during the charging and discharging processes, respectively.

During the charging and discharging processes, the power should be limited, as follows:

$$0 \leq P_{B,c}(t) \leq P_{B,c,\max} \tag{29}$$

$$P_B(t) = P_{B,c}(t) - P_{B,d}(t) \tag{30}$$

In our work, we assumed that the energy stored in the batteries during the end scheduling period is greater than $S_{Bi,base}$, to ensure that the batteries have stored energy available for the next day. This limit can be expressed as

$$S_{Bi}^{*} \geq S_{Bi,base} \tag{31}$$

where S_{Bi}^{*} is the energy of the ith BESS in the end dispatch period; $S_{Bi,base}$ is the minimum required dispatched energy of the ith BESS.

The power balance equation of the MG can therefore be modified as follows:

$$\sum_{i=1}^{N_G} P_{G_m i}(t) + \sum_{i=1}^{N_B} P_{Bi}(t) + \sum_{i=1}^{N_{WT}} P_{WTi}(t) + \sum_{i=1}^{N_{PV}} P_{PVi}(t) + P_{Tie}(t) = P_{load} \tag{32}$$

The inequality constraint of the MT output limit can be expressed as

$$P_{G_m i,\min}(t) \leq P_{G_m i}(t) \leq P_{G_m i,\max}(t) \tag{33}$$

4. Optimal Solution

In bi-level optimization problems, the decision variables of the upper-level are taken as the parameters in the lower-level. Here, the exchanged power between the DSO and MGO applied the energy band for the tie-line flow, which is the upper-level variable, is considered as the parameter for the lower-level. In our work, instead of transferring the bi-level optimization problem into a single-level, a multi-objective optimization problem relating to the cooperation between the DSO and MGO is applied, in which the modified NSGA-II (MNSGA-II) is used to solve the proposed model and applied to the cooperation of the variable relationships between the DSO and MGO. Moreover, the controlled elitism and dynamic crowded tournament selection have been applied for criteria of Pareto-optimal by using the MNSGA-II. Then, the tie-line constraints convergence criterion is solved in the multi-objective problem form checking the post state feasibility. If the tie-line constraints convergence criterion is satisfied, the optimization process can proceed to the next step. If not, the iteration number can increase, and the above process repeated. Owing to the trade-off between the stable operation of the DN and economical operation of the MG, the DSO and MGO cautiously consider the "Pareto solution" to solve the multi-objective problem with bi-level optimization while improving the entire system operations.

4.1. NNSGA-II

The conventional NSGA-II includes two principal components: a non-dominated sorting solution and crowding distance (CD) sorting procedure for preserving the solution diversity [30]. The NSGA-II uses crossover and mutation operators to create the offspring population and adopts a rapid non-dominated sorting method to decide the non-dominated rank of individuals. Since all members of the previous solution injecting the new population may not be compiled, only several individuals corresponding to the number of available fronts can be chosen from the last solution based on the CD. Parents are also picked from the population using the crowded tournament selection method based on the rank and CD. The crowded tournament selection in the NSGA-II randomly identifies any two objects and selects the one in the less crowded region, when the two objects have the same non-domination level. The object with the lower rank or higher CD is decided. The adopted population makes offspring according to the crossover and mutation operators. Furthermore, this algorithm employs an elite preservation strategy to choose the new generation from the parent and offspring population. The CD sorting procedure calculates the dispersion of the solutions in each solution and maintains the Pareto solution diversity.

$$CD_i = \frac{1}{N_{obj}} \sum_{g=1}^{N_{obj}} \left| f_{i+1}^g - f_{i-1}^g \right| \tag{34}$$

where f_{i+1}^g and f_{i-1}^g are the gth objective of the $i + 1$th and $i - 1$th individuals, respectively.

Although the NSGA-II contains improved concepts such as elitism, rapid non-dominated sorting, and diversity maintenance along the Pareto-optimal solution, it remains insufficient with respect to preserving the lateral diversity and a uniform distribution of the non-dominated solutions. An emphasis on lateral diversity is necessary to prevent excessive exploitation and thereby ensures that the search algorithm converges more rapidly. A stable distribution of non-dominated solutions is necessary to include the optimal Pareto solutions. To address the disadvantages of the NSGA-II, controlled elitism and the dynamic CD (DCD) are applied as the criteria for the optimal Pareto solution [31]. Therefore, in this study, it was ensured that the criterion for the multi-objective optimization process was satisfied by the convergence process, based on the criteria for assessing the Pareto optimal solution, which involves the controlled elitism and DCD of the proposed MNSGA-II.

$$DCD_i = \frac{CD_i}{log\left(\frac{1}{Var_i}\right)} \tag{35}$$

where Var_i is the variance of CDs calculated by Equation (36). Var_i is based on

$$Var_i = \frac{1}{N_{obj}} \sum_{g}^{N_{obj}} \left(\left| f_{i+1}^g - f_{i-1}^g \right| - CD_i \right)^2 \tag{36}$$

Regarding controlled elitism, the MNSGA-II regulates the number of objects in the optimal selection adaptively and preserves a pre-defined number of distributed objects in each solution. Firstly, the integrated parent and offspring population $R_h = Pop_h \cup Off_h$ is divided for non-domination. Let Nf be the number of non-dominated solutions in the incorporated population (of size 2 M). According to the geometric distribution, the maximum available number of objects decided in the yth case ($y = 1, 2, \ldots, Nf$) in the new population of size M_y is expressed:

$$M_y = M \cdot \frac{1 - \gamma}{1 - \gamma^{Nf}} \cdot \gamma^{y-1} \tag{37}$$

where M_y is the new population size; M is the population size; γ is the reduction rate.

Since $\gamma < 1$, the maximum available number of objects in the solution is the highest, and other solutions are permitted to contain an exponentially decreasing number of solutions.

4.2. Solution Procedure

The bi-level optimization for cooperation between the DSO and MGO is implemented in the following sequential manner:

Step 1. Set the input parameters as well as the lower and upper limits of each power system variable for the bi-level optimization process.

Step 2. Choose the population size M, crossover and mutation probability, crossover and mutation index, and maximum number of generations.

Step 3. Solve the upper-level without the lower-level and obtain the initial base-case solution.

Step 4. After obtaining the base-case solution by solving the upper-level without any energy band constraints, solve the lower-level to determine the new base case. Here, the tie-line flow from the energy band between the DSO and MGO is taken as a decision variable, and this variable is returned to the upper-level.

Step 5. When violations are detected in the lower-level, solve the upper-level with all of the feedback information included, thereby creating a new base-case.

Step 6. Solve the lower-level in parallel with the new base-case.

Step 7. Repeat the bi-level process until the new base-case is established.

Step 8. If the multi-objective functions are to obtain the converged Pareto solution, the process terminates; otherwise, it is repeated from Step 3.

A flowchart of the detailed approach is presented in Figure 3.

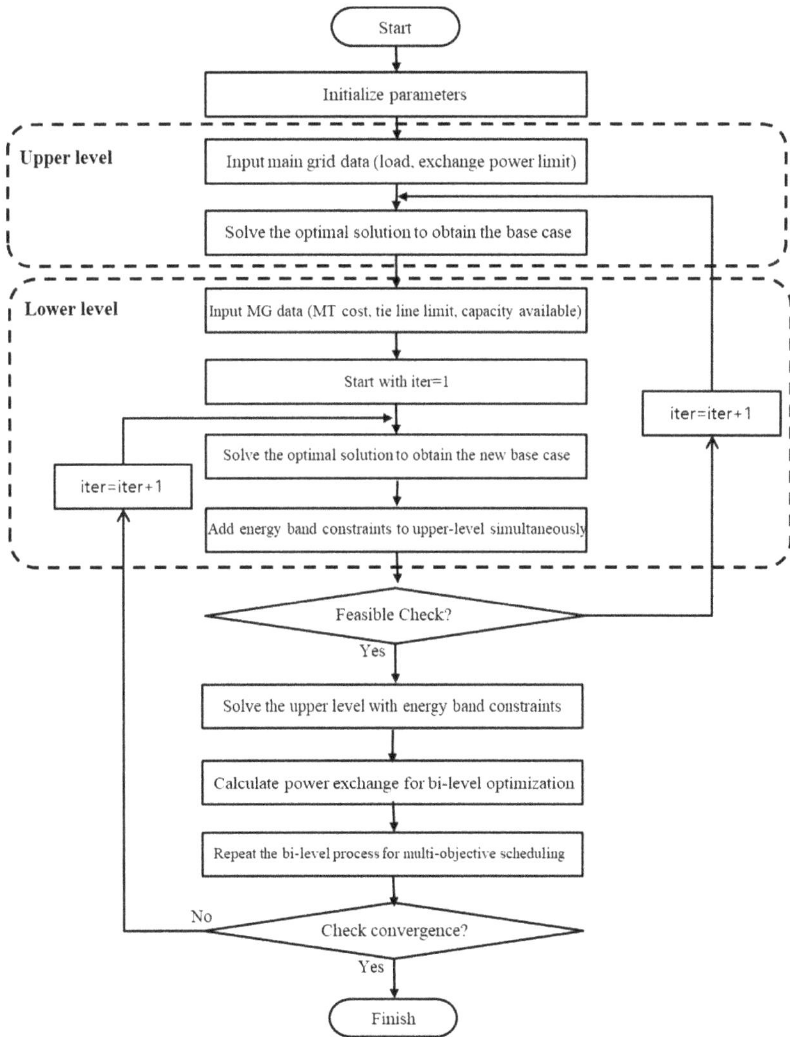

Figure 3. Bi-level optimization process.

5. Result and Discussion

5.1. Data

The MG contains distributed generation units, including WTs, MTs, and PV panels. The power for the WTs, PV panels, and total load were taken from [28], as shown in Figure 4. Since the load amount is greater than the total renewable energy at most times, other DER sources such as the MT and BESS are necessary to balance the MG demand. In this paper, we assume that the three types of loads (industrial, commercial, and residential customers) have a common characteristic and can be considered as critical and interruptible loads. Therefore, the three types of loads can be added together and considered the total load. Here, the total peak load is 500 kW. The MG structure is equipped with WT whose total installed capacity is 500 kW, PV 300 kW, and MT 500 kW.

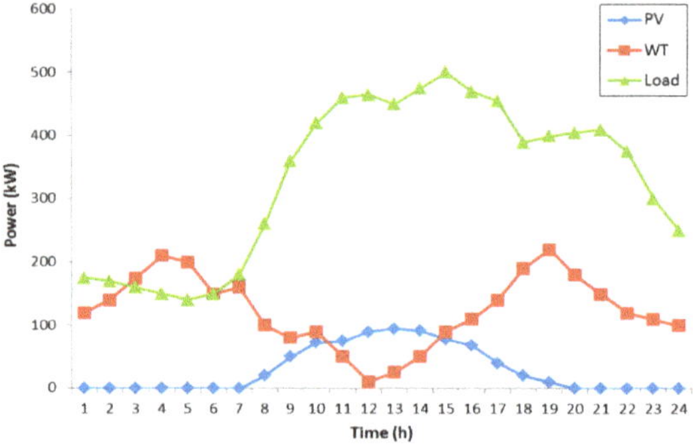

Figure 4. Forecasting power of PV panels and WTs and load for test system.

In this study, the capacity of the BESS was 200 kWh, while the charging and discharging ramp-rate limits were both 50 kW/h. The battery efficiencies were 0.9 at any time step during the charging and discharging processes [32]. Furthermore, the renewable energy in the operating cost was assumed to have little effect on the final result. It was also assumed that no energy exchange occurred between MGs. All characteristics of the DERs and other values of the technical parameter are depicted in Table 1. Since we consider not only the scheduling in the MG, but also the operation of the tie-line, we assume that the minimum rated capacity of MT is set to zero in order to clarify the tie-line operation strategy considering the energy band. Figure 5 represents the forecasted real-time pricing (RTP) of the wholesale market and the retail market prices based on the time-of-use (TOU) scheme [33].

Table 1. Values of technical parameters.

Parameter	Value	Parameter	Value
ρ_{loss} ($/kWh)	250	ρ_{PVi} ($/kWh)	11
$P_{M,max}$ (kW)	36	ρ_{BESSi} ($/kWh)	8
$P_{e,max}$ (kW)	20	V_{min}, V_{max} (p.u.)	0.95, 1.05
ρ_{mi} ($/kWh)	71	$P_{Gmi,min}$, $P_{Gmi,max}$ (kW)	0, 20
$P_{WT,min}$, $P_{WT,max}$ (kW)	0, 250	$P_{PV,min}$, $P_{PV,max}$ (kW)	0, 100

The residual energy curves of the BESS for a period of 24 h are shown in Figure 6. The initial state of charge of the BESS was 90 kWh. At 08:00, the BESS was fully charged. During the high-price period, the BESS injected power into the MG. Furthermore, at the end of the day, the residual energy of the BESS decreased to the initial value, which means that the energy of the BESS remained balanced throughout the day.

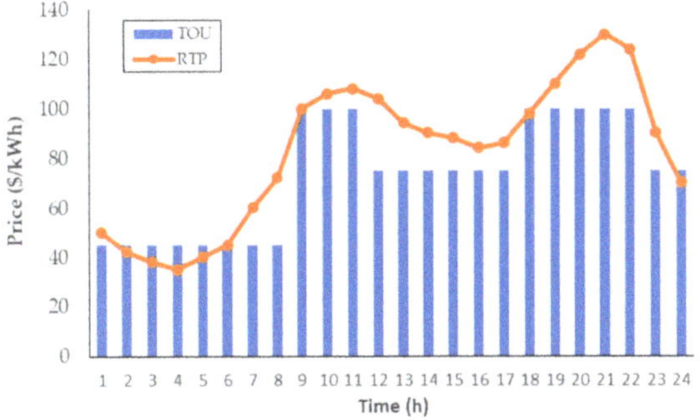

Figure 5. Market price information.

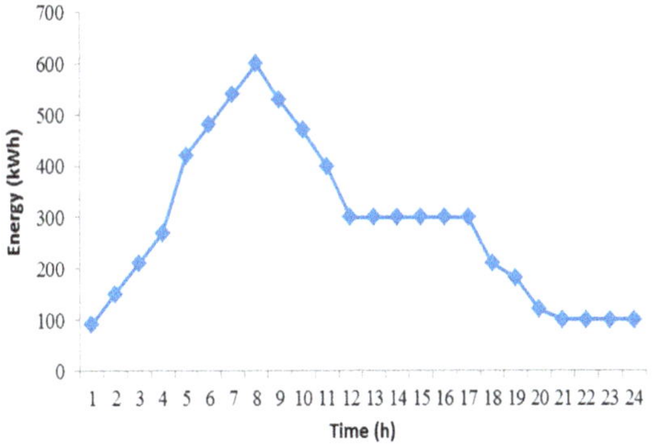

Figure 6. Residual energy of BESS.

5.2. Simulation Results

To evaluate the superior performance of the proposed operation scheme, the following two cases were considered: Case 1: a bi-level optimization model that considers the tie-line flow without the energy band and Case 2: a bi-level optimization model that considers the tie-line flow within the energy band.

5.2.1. IEEE Test System

To demonstrate the validity of the proposed approach, an IEEE test system in which general European MGs [34] connected the modified IEEE 33-bus DN [35] was used for the simulation and analysis is shown in Figure 7. Details about the IEEE 33-bus DN can be found in [36]. The factors should be taken into account when modeling an MG integrated with the DN. Here, the MG was always connected to the DN by means of the tie-line. The maximum limit of the exchange power of the grid tie-line was 400 kW. The scheduling period was assumed to be a single day.

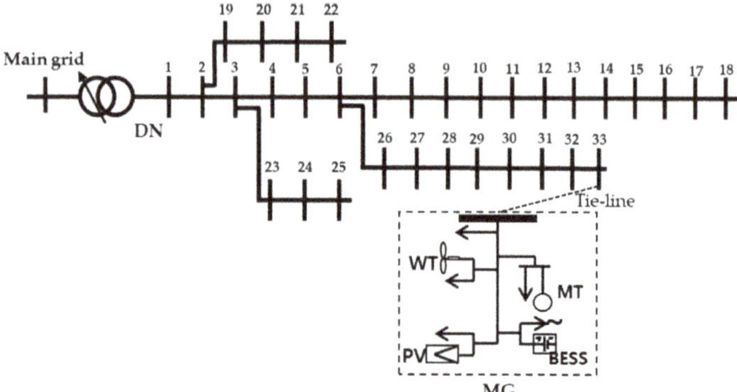

Figure 7. IEEE (Institute of Electrical and Electronics Engineers) test system.

Figure 8 illustrates the resulting scheduled power scheme for the distributed units of the integrated MG, when the energy band was not considered, where the tie-line power accounted for the majority of the load. Moreover, the BESS charged and discharged less frequently. The batteries were charged from 02:00 to 04:00, while they were discharged at 07:00 and 19:00. Furthermore, the MG purchased energy from the DN to optimize the operational cost, because it was significantly lower than the MT cost. The manner in which the MGO shares the responsibility for ensuring balance of the supply and demand with the DSO under the fluctuating renewable energy depends on the operating condition of the tie-line flow.

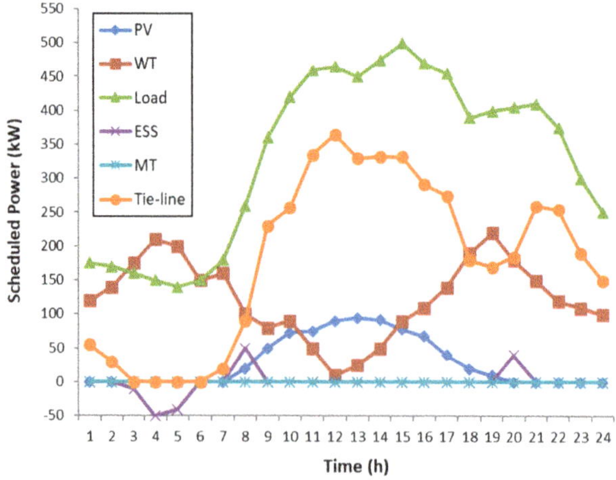

Figure 8. Scheduled power scheme for distributed units for Case 1.

The tie-line flow results of Case 1 are based on the predetermined (contractual) tie-line flow scenario. These results demonstrate that the MG rescheduled receiving the control signal from the DN for operating cost minimization with no tie-line constraints. Interestingly, the tie-line flow changed significantly to approximately 140 kW from 08:00 to 09:00, with a corresponding change in the load of 360 kW. Such substantial changes involve an increase in the potential risks or operational costs of the DN, even though this results in significant reductions in the operating costs of the MG. The important issue is the reduction of the reliability for the entire DN, which is far more significant than the minimization in the operating costs of the MG. This means that the scalability, namely the ability to incorporate MGs into the DN, may be restricted. In this case, fluctuations as large as 4334 kW could occur in the power transmission via the tie-line during the day.

We also evaluated the energy band operation scheme where the MG optimized the control signal of the tie-line flow within the energy band (Case 2). The sharing of reliability should be carefully considered when establishing the energy band size as an operational condition, owing to the trade-off between reliability and cost in the grid-connected system. We considered energy bands ranging from 30% to 80% and compared the results with those of Case 1, as indicated in Table 2. According to the results, the ratio of the 80% energy band was almost equal to the contract power. In the proposed bi-level optimization model, ensuring the proposed ramping capability in the MG is essential for mitigating operational risks.

Table 2. Comparison of MG operational costs.

Time (h)	Contract (kW)	Scheduled Power (kW) (Energy Band)					
		30%	40%	50%	60%	70%	80%
1	55	40.04	43.12	46.2	49.28	52.36	55.44
2	30	21.06	22.68	24.3	25.92	27.54	29.16
3	0	0	0	0	0	0	0
4	0	0	0	0	0	0	0
5	0	0	0	0	0	0	0
6	0	0	0	0	0	0	0
7	20	14.04	15.12	16.20	17.28	18.36	19.44
8	90	63.18	68.04	72.90	77.76	82.62	87.48
9	230	161.46	173.88	186.30	198.72	211.14	223.56
10	257	180.41	194.29	208.17	222.05	235.93	249.80
11	335	235.17	253.26	271.35	289.44	307.53	325.62
12	365	256.23	275.94	295.65	315.36	335.07	354.78
13	330	231.66	249.48	267.30	285.12	302.94	320.76
14	333	233.77	251.75	269.73	287.71	305.69	323.68
15	332	233.06	250.99	268.92	286.85	304.78	322.70
16	292	204.98	220.75	236.52	252.29	268.06	283.82
17	275	193.05	207.90	222.75	237.60	252.45	267.30
18	180	126.36	136.08	145.80	155.52	165.24	174.96
19	170	119.34	128.52	137.70	146.88	156.06	165.24
20	185	129.87	139.86	149.85	159.84	169.83	179.82
21	260	182.52	196.56	210.60	224.64	238.68	252.72
22	255	179.01	192.78	206.55	220.32	234.09	247.86
23	190	133.38	143.64	153.90	164.16	174.42	184.68
24	150	105.30	113.40	121.50	129.60	137.70	145.80
Ratio		70.23	75.64	81.04.	86.44	91.84	97.25

The scheduled power scheme for Case 2 with an energy band of 80% is plotted in Figure 9. It can be observed from Figure 9a that the scheduled power scheme was similar to the contract power. Limited control over the tie-line flow could cause the power system reliability of the MG to decrease owing to an increased supply–demand unbalance. As shown in Figure 9b, the BESS was charged for longer than that during Case 1 (from 03:00 to 05:00). The MT also generated power, which was used to ensure reliable operation between the DSO and MGO, because of the ramping capability of the proposed energy band scheme. Consequently, these results demonstrate that the increased responsibility for balancing the supply and demand requires preparation and planning to make sure that the MGO is capable of ramping.

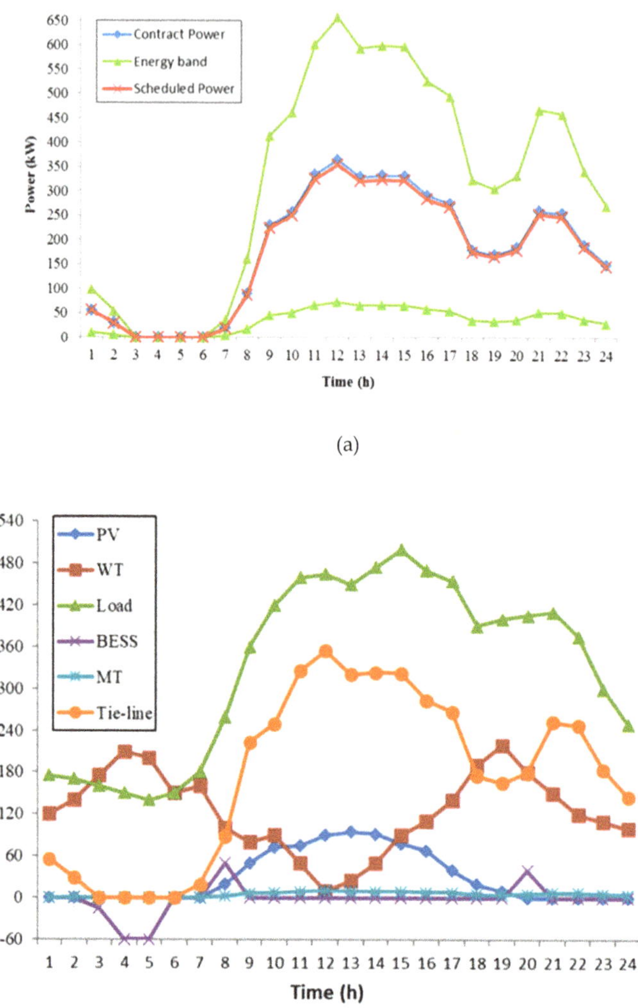

(a)

(b)

Figure 9. Scheduled power scheme for Case 2 for energy band of 80%: (**a**) Results for energy band, (**b**) Scheduled power volumes for distributed units.

Table 3 compares the operational costs of the DN for all of the cases. In the base case, that is, when no power was purchased from the DN, the operational cost was $519,482/h, while the operational costs for Cases 1 and 2 were $516,841/h and $517,212/h, respectively. Although the operational cost in the case of the proposed bi-level optimization model (Case 2) was 0.2% ($1033/h) higher than that for Case 1, it was 0.3% ($1608/h) lower than that for the base case. These results are relevant to the problems relating the MG operation; that is preventing shortages in the energy supply to the DN from worsening, even though the operational costs for both Case 2 and the base case were higher than that for Case 1.

Table 3. Comparison of the operational costs for each case.

Case	Operation Cost ($/h)		Note
	DSO	MGO	
Base	856,374	519,482	No purchase of power from DN
Case 1	873,622	516,841	Purchase of power from DN through tie-line
Case 2	862,716	517.874	Purchase of power from DN with energy band of 80%

To demonstrate the effectiveness of the bi-level optimization model for cooperation between the DSO and MGO, the following two scenarios of Case 2 were considered: Scenario A: no bi-level optimization model with Case 2, and Scenario B: bi-level optimization model with Case 2. Table 4 presents a comparison of the operational costs for each scenario in Case 2. In Scenario A, the MG purchased power from the wholesale market at the RTP, because the feedback information including the operational costs of MG and energy band, was not considered. According to the comparison results, the operational cost of the MG in Scenario A was $512,638/h, which was lower than that in Scenario B, namely $517,874/h, but the operational cost of the DN in Scenario A was higher than that in Scenario B.

Table 4. Comparison of operational costs for scenarios in Case 2.

Scenario	Operation Cost ($/h)		Note
	DSO	MGO	
Scenario A	864,274	512,638	No bi-level model with Case 2
Scenario B	862,716	517,874	Proposed bi-level model with Case 2

The optimal results of the proposed approach for each case are summarized in Table 5. Comparing the operation cost of the DSO, Case 2 was lower than Case 1, while the MGO profit in Case 2 was lower than Case 1. It can be observed that the DSO operational cost was the lowest, even though the MGO profit was little reduced in Case 2. Therefore, the results indicated in our study confirm that the proposed approach provides the DSO and MGO with the optimal solution for reliable and economical operation.

Table 5. Comparison of the optimal results for each case.

Case	Operational Cost (DSO)	Profit (MGO)
Case 1	873,622	747,216
Case 2	862,716	746,085

5.2.2. Actual Power System

We evaluated the practical applicability of the proposed approach for large-scale power systems, by applying the results of our work to an actual power system in Shandong, China, called the Changdao project [37], consisting of 128 buses and 7 MGs as shown in Figure 10. All of the MGs were constantly connected to the DN by means of a single tie-line. The maximum limit of the tie-line flow was 1000 kW. The scheduling period was considered to be a single day. The data for every MG and the capacity parameters were listed in Table 6.

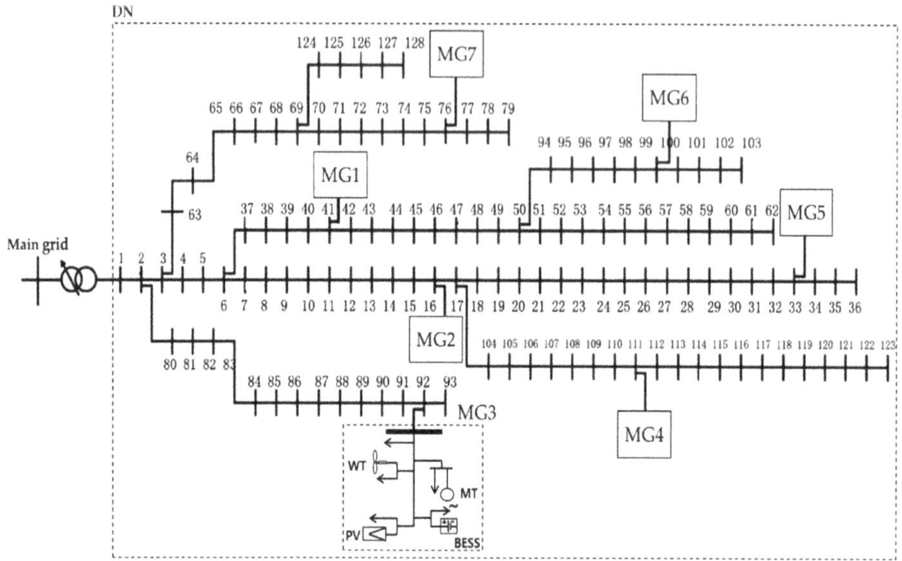

Figure 10. Actual power system in China.

Table 6. Capacity parameters.

Type	Minimum Power (kW)	Maximum Power (kW)	Number of Units	Ramp up/down Rate (kW/min)
MT	0	1000	11	500/400
WT	0	1000	11	-
PV	0	1500	10	-
ESS	−1500	1500	7	-

The scheduled power scheme for Case 2 with an energy band of 60% is shown in Figure 11. As can be seen from Figure 11b, the BESS was charged for longer, from 04:00 to 07:00, than in Case 1, with the MTs also generating additional power. These results indicate that the proposed bi-level optimization model with the energy band can effectively improve reliable operations between the DSO and MGO.

Table 7 compares the operational costs for each case. In the base case, the operational cost was $1,454,549/h when no power was purchased from the DN. In comparison, the operational costs for Cases 1 and 2 were $1,447,154/h and $1,450,047/h, respectively. Although the operational cost for the proposed scheme (Case 2) was 0.2% ($2893/h) higher than that for Case 1, it was lower than that for the case in which no power was purchased from the DN by approximately 0.3% ($4532/h). Therefore, the proposed bi-level optimization model based on a 60% energy band provides superior balance for economical and reliable operation between the DSO and MGO.

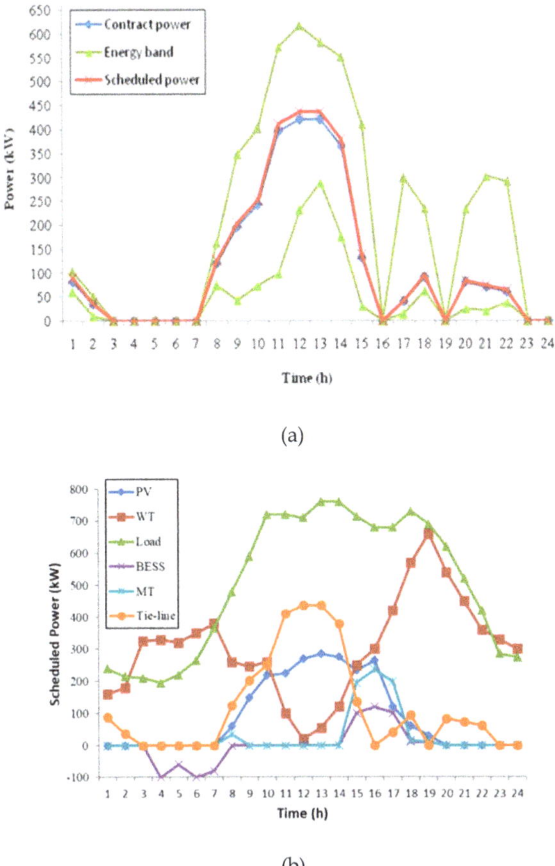

(a)

(b)

Figure 11. Scheduled power scheme for Case 2 within energy band of 60%: (**a**) Results for energy band, (**b**) Scheduled power volumes for distributed units.

Table 7. Comparison of operational costs for each case.

Case	Operation Cost ($/h)		Note
	DSO	**MGO**	
Base	2,081,643	1,454,549	No purchase of power from DN
Case 1	2,548,231	1,447,154	Purchase of power from DN through tie-line
Case 2	2,320,075	1,450,047	Purchase of power from DN within energy band of 60%

Table 8 presented a comparison of the operational costs for each scenario in Case 2. To confirm the results, we assumed the same scenario as in the IEEE test system described previously. According to the comparison results, the MG operational cost for Scenario A was $1,387,562/h, which was lower than that for Scenario B, namely $1,450,047/h, but the operational cost of the DSO for Scenario A is higher than that for Scenario B. It can be observed that the proposed bi-level model can reduce the DSO operational cost although the MGO operational cost increases, because the MG operational cost can be reduced by the proposed tie-line control strategy.

Table 8. Comparison of operational costs for each Scenario in Case 2.

Scenario	Operation Cost ($/h)		Note
	DSO	**MGO**	
Scenario A	2,527,983	1,387,562	No bi-level model with Case 2
Scenario B	2,320,075	1,450,047	Proposed bi-level model with Case 2

Table 9 displayed the optimal results of the proposed approach for each case. Similar to the result for the IEEE test system previously, the operation cost of the DSO in Case 2 was lower than Case 1, while the MGO profit in Case 2 was lower than Case 1. It can be seen that the DSO operational cost was the lowest, even though the MGO profit was little reduced in Case 2. Therefore, it should be noted that the proposed approach provides the optimal balance for reliable and economical operation between the DSO and MGO.

Table 9. Comparison optimal results for each case.

Case	Operational Cost (DSO)	Profit (MGO)
Case 1	2,548,231	988,274
Case 2	2,320,075	980,610

5.3. Performance Test

The multi-objective performance of the MNSGA-II was evaluated by comparing the Pareto solutions obtained using the NSGA-II and MNSGA-II for 30 simulations performed on both the IEEE test system and actual power system. The convergence metric, spread/diversity metric, inverted generational distance, and minimum spacing metric were calculated for the non-dominated solutions obtained using the NSGA-II and MNSGA-II [38]. A total of 30 independent trials were conducted to select the appropriate values for these parameters and the optimal parameters selected are presented in Table 10. In this case, the initial weight factor, w, was set to 0.5.

Table 10. Parameters used for non-dominated sorting genetic algorithm II (NSGA-II) and modified version of the NSGA-II (MNSGA-II).

Parameter	IEEE Test System		Actual Power System	
	NSGA-II	**MNSGA-II**	**NSGA-II**	**MNSGA-II**
Population size	100	100	200	200
Max. no. of generations	30	30	30	30
Crossover probability	0.8	0.8	0.9	0.9
Mutation probability	1/12	1/12	1/75	1/75
Crossover index	1	1	2	2
Mutation index	10	10	20	20

Figure 12 displays the optimization results of the Pareto solution obtained in Case 2 among the particle swarm optimization (PSO) [39], adaptive modified PSO (AMPSO) [40], strength Pareto evolutionary algorithm (SPEA) [41], genetic algorithm (GA) [42], NSGA-II [43], and MNSGA-II on the test system and actual system. Note that the third point converged as the Pareto solution in the case of MNSGA-II for both systems. Table 11 indicates the comparison results for each system. According to the comparison results, it is clear that the DSO operating cost and MGO profit were lower in the proposed MNSGA-II than in the other algorithms for both systems. These results may be small in terms of the overall operation, but a value is not small in terms of the cost regarding the margin of the tie-line between the DN and MG for sharing the exchange information. As shown in Table 11, the run time of the proposed approach was significantly reduced compared to the others, particularly for large power systems, owing to the parallel processing based on the bi-level model. These results demonstrate that the proposed approach is suitable for the requirements of realistic power system operation.

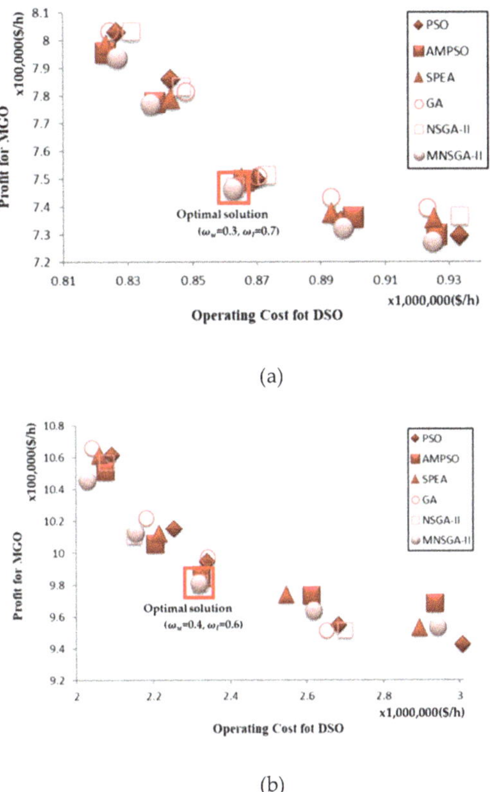

Figure 12. Optimization results for each system. PSO: particle swarm optimization; adaptive modified PSO; SPEA: strength Pareto evolutionary algorithm; GA: strength Pareto evolutionary algorithm; NSGA-II: non-dominated sorting genetic algorithm II; MNSGA-II: modified version of the non-dominated sorting genetic algorithm II: (**a**) IEEE test system, (**b**) Actual power system.

Table 11. Comparison result of different algorithms for Case 2.

Test System	Algorithm	DSO Operating Cost ($/h)	MGO Profit ($/h)	Total Run Time (min)
IEEE test system	PSO	869,694	749,964	6.27
	AMPSO	866,047	749,165	1.95
	SPEA	865,147	750,132	2.03
	GA	870,486	750,641	5.83
	NSGA-II	863,208	746,927	8.54
	MNSGA-II	862,716	746,085	1.83
Actual power system	PSO	2,339,648	994,648	22.74
	AMPSO	2,325,348	985,912	8.97
	SPEA	2,328,321	984,315	10.84
	GA	2,341,934	997,324	21.25
	NSGA-II	2,322,496	981,073	24.31.
	MNSGA-II	2,320,075	980,610	5.46

The superior performance of the optimization process for multi-objective problem obtained Pareto-fronts between DSO and MGO using the MNSGA-II are shown in Figure 13. As shown in Figure 13a, when the weighting factors were w_u = 0.3 (upper-level) and w_l = 0.7 (lower-level), the convergence points of the Pareto-front for profit and operating cost in the IEEE test system were obtained after the 21th iterative cycle. On the other hand, when the weighting factors were w_u = 0.4 and w_l = 0.6, the Pareto-front converged after the 28th iterative cycle (see Figure 13b). On the basis of the results in Figure 13, it can be concluded that the performance of multi-objective optimization was superior, as it exhibited higher convergence speed and required fewer iterative cycles to converge. Therefore, the proposed approach results in optimal operational solutions and can assist in decision-making whenever there is a trade-off between the DSO and MGO in power systems.

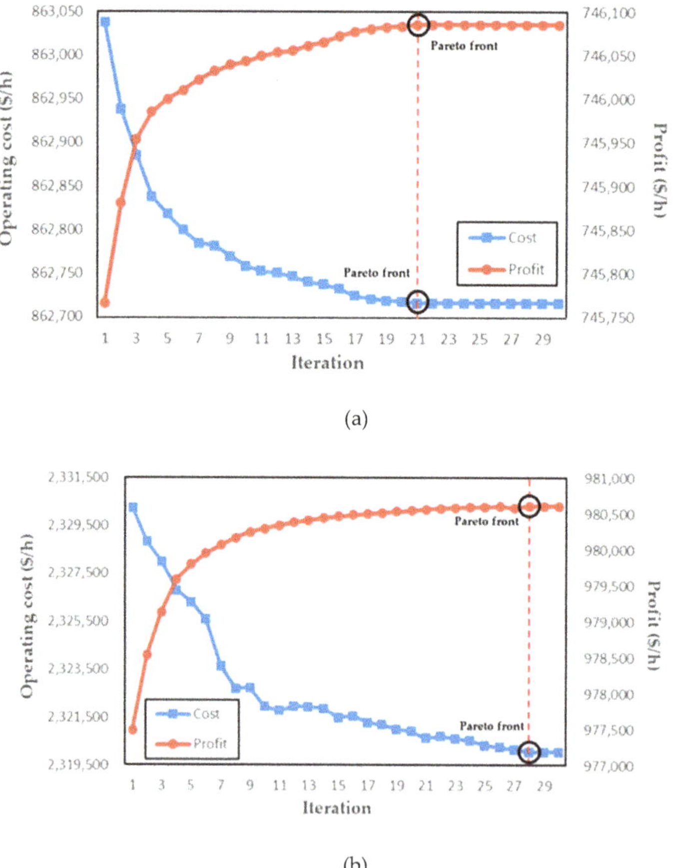

Figure 13. Optimization process for obtained Pareto-front via MNSGA-II: (**a**) IEEE test system, (**b**) Actual power system.

6. Conclusions

A bi-level operation model based on the energy band has been proposed for improved cooperation. The upper-level model describes the optimal dispatch of the DN with the aim of minimizing the operating cost. The lower-level model considers the scheduling information obtained by the upper-level for feedback information as the objective for maximizing the MG profit for cooperation between the

DSO and MGO. In our work, instead of converting the bi-level optimization problem into a single level, a multi-objective optimization problem relating to the cooperation between the DSO and MGO was presented, in which the MNSGA-II was used to solve the proposed model, and applied to the cooperation of the variable relationships between the DSO and MGO. Owing to the trade-off between the stable operation of the DN and economical operation of the MG, the DSO and MGO cautiously consider the "Pareto solution" to solve the multi-objective problem with bi-level optimization while improving the operations of the entire system. The simulation results demonstrated that when an MG is integrated into the DN, the benefits of the entire system are optimized. During the simulations, by using this approach, the tie-line flow could be managed smoothly within the energy band, although the results for the grid-connected MG operation-related approach showed that the tie-line flow changes sharply. Therefore, the proposed optimization approach provides additional economic benefits for power systems, along with performance improvements and increased reliability for cooperation between the DSO and MGO. Our future work will focus on the reasonable selection of additional energy band values while considering various factors as well as the tie-line flow.

Author Contributions: H.-Y.K. proposed the main idea of this paper and M.-K.K. coordinated the proposed approach and thoroughly reviewed the manuscript. H.-J.K. provided essential information and supported manuscript preparation. All authors read and approved the manuscript.

Funding: This research received no external funding.

Acknowledgments: This research was supported by the Chung-Ang University Research Scholarship Grants in 2019. This research was also supported by the Korea Electric Power Corporation (Grant number: R18XA06-75).

References

1. Kim, H.Y.; Kim, M.K. Optimal generation rescheduling for meshed AC/HIS grids with multi-terminal voltage source converter high voltage direct current and battery energy storage system. *Energy* **2017**, *119*, 309–321. [CrossRef]
2. Sheikhahmadi, P.; Mafakheri, R.; Bahramara, S.; Damavandi, M.Y.; Catalao, J. Risk-based two-stage stochastic optimization problem of micro-grid operation with renewables and incentive based demand response programs. *Energies* **2018**, *11*, 610. [CrossRef]
3. Zamora, R.; Srivastava, A.K. Controls for microgrids with storage review, challenges, and research needs. *Renew. Sustain. Energy Rev.* **2010**, *14*, 2009–2018. [CrossRef]
4. Shi, L.; Luo, Y.; Gy, T. Bidding strategy of microgrid with consideration of uncertainty for participating in power market. *Int. J. Electr. Power Energy Syst.* **2014**, *59*, 1–13. [CrossRef]
5. Feijoo, F.; Das, T.K. Emissions control via carbon policies and microgrid generation: A bilevel model and pareto analysis. *Energy* **2015**, *90*, 1545–1555. [CrossRef]
6. Kim, D.; Kwon, H.G.; Kim, M.K.; Park, J.K.; Park, H.G. Determining the flexible ramping capacity of electric vehicles to enhance locational flexibility. *Energies* **2017**, *10*, 2028. [CrossRef]
7. Erdine, O. Economic impacts of small-scale own generating and storage units, and electric vehicles under different demand response strategies for smart households. *Appl. Energy* **2014**, *126*, 142–150. [CrossRef]
8. Akorede, M.F.; Hizam, H.; Pouresmaeil, E. Distributed energy resources and benefits to the environment. *Renew. Sustain. Energy Rev.* **2010**, *14*, 724–734. [CrossRef]
9. Talari, S.; Khah, M.S.; Osorio, G.; Aghael, J. Stochastic modeliling of renewable energy sources from operatiors' point of view: A survey. *Renew. Sustain. Energy Rev.* **2017**, *81*, 1953–1965. [CrossRef]
10. Kou, P.; Feng, Y.; Liang, D.; Gao, L. A model predictive control approach for matching uncertain wind generation with PEV charging demand in a microgrid. *Int. J. Electr. Power Energy Syst.* **2019**, *105*, 488–499. [CrossRef]
11. Rist, J.F.; Dias, M.F.; Palman, M.; Zelazo, D.; Cukurel, B. Economic dispatch of a single micro-gas turbine under CHP operation. *Appl. Energy* **2017**, *200*, 1–18. [CrossRef]
12. Velik, R.; Nicolay, P. Grid price dependent energy management in microgrid using a modified simulated annealing triple optimizer. *Appl. Energy* **2014**, *130*, 384–395. [CrossRef]

13. Palizban, O.; Kauhaniemi, K.; Guerrero, J.M. Microgrids in active network management—Part II: System operation, power quality and protection. *Renew. Sustain. Energy Rev.* **2014**, *36*, 440–451. [CrossRef]
14. Sechilariu, M.; Wang, B.C.; Locment, F.; Jouglet, A. DC microgrid power flow optimization by multi-layer supervision control, design and experimental validation. *Energy Convers. Manag.* **2014**, *82*, 1–10. [CrossRef]
15. Sechilariu, M.; Wang, B.C.; Locment, F. Supervision control for optimal energy cost management in DC microgrid: Design and simulation. *Int. J. Electr. Power Energy Syst.* **2014**, *58*, 140–149. [CrossRef]
16. Li, T.S.; Zhang, H.G.; Huang, B.N.; Teng, F. Distributed optimal economic dispatch based on multi-agent system framework in combined heat and power systems. *Appl. Sci.* **2016**, *6*, 308. [CrossRef]
17. Jiang, Q.; Xue, M.; Geng, G. Energy management of microgrid in grid connected and stand-alone modes. *IEEE Trans. Power Syst.* **2013**, *28*, 3380–3389. [CrossRef]
18. Liu, G.; Mahmoudi, N.; Chen, K. Microgrids real-time pricing based on clustering techniques. *Energies* **2018**, *11*, 1388. [CrossRef]
19. Tan, S.; Xu, J.; Panda, S.K. Optimization of distribution network incorporating distributed generators: An integrated approach. *IEEE Trans. Power Syst.* **2013**, *28*, 2421–2432. [CrossRef]
20. Khodaei, A.; Shahidehpour, M. Microgrid-based co-optimization of generation and transmission planning in power systems. *IEEE Trans. Power Syst.* **2013**, *28*, 1582–1590. [CrossRef]
21. Zenginis, I.; Vardakas, J.S.; Echave, C.; Morato, M.; Abadal, J. Cooperation in microgrids through power exchange: An optimal sizing and operation approach. *Appl. Sci.* **2017**, *203*, 972. [CrossRef]
22. Bahramara, S.; Moghaddam, M.P.; Haghifam, M.R. A bi-level optimization model for operation of distribution networks with micro grids. *Int. J. Electr. Power Energy Syst.* **2016**, *82*, 169–178. [CrossRef]
23. Shi, N.; Luo, Y. Bi-level programming approach for the optimal allocation of energy storage systems in distribution networks. *Appl. Sci.* **2017**, *7*, 398. [CrossRef]
24. Xie, J.; Zhong, J.; Li, Z.; Gan, D. Environmental economic unit commitment using mixed integer linear programming. *Eur. Trans. Electr. Power* **2011**, *21*, 772–786. [CrossRef]
25. Muthuswamy, R.; Krishnan, M.; Subramanian, K.; Subramanian, B. Environmental and economic power dispatch of thermal generators using modified NSGA-II algorithm. *Int. Trans. Electr. Energy Syst.* **2014**, *25*, 1552–1569. [CrossRef]
26. Lee, S.Y.; Jin, Y.G.; Yoon, Y.T. Determining the optimal reserve capacity in a microgrid with islanded operation. *IEEE Trans. Power Syst.* **2016**, *31*, 1369–1376. [CrossRef]
27. Colson, B.; Marcotte, P.; Savard, G. An overview of bilevel optimization. *Ann. Oper. Res.* **2007**, *153*, 235–256. [CrossRef]
28. Liu, Y.; Jiang, C.; Shen, J.; Zhou, X. Energy management for grid-connected micro grid with renewable energies and dispatched loads. *Prz. Elektrotechniczny Electr. Rev.* **2012**, *88*, 87–93.
29. Mazidi, M.; Zakariazadeh, A.; Jadid, S.; Siano, P. Integrated scheduling of renewable generation and demand response programs in a microgrid. *Energy Conv. Manag.* **2014**, *86*, 1118–1127. [CrossRef]
30. Deb, K. *Multi-Objective Optimization Using Evolutionary Algorithms*; Wiley: Chichester, UK, 2001.
31. Luo, B.; Zheng, J.; Xie, J.; Wu, J. Dynamic crowding distance—A new diversity maintenance strategy for MOEAs. In Proceedings of the IEEE International Conference on Natural Computation, Jinan, China, 18–20 October 2008; pp. 580–585.
32. Mingrui, Z.; Jie, C.; Zhichao, D.; Shaobo, W.; Hua, S. Economic operation of microgrid considering regulation of interactive power. *Chin. Soc. Electr. Eng.* **2014**, *34*, 1013–1023.
33. Sahedi, M.M.; Duki, E.A.; Kia, M. Simultanous emergency demand response programming and unit commitment programming in comparison with interruptible load contracts. *IET Gener. Transm. Distrib.* **2012**, *6*, 605–611.
34. Papathanassiou, S.; Hatziargyrlou, N.D.; Strunz, K. A benchmark low voltage microgrid for steady state and transient analysis. In Proceedings of the CIGRE Symposium: Power Systems with Dispersed Generation, Athens, Greek, April 2005.
35. Wang, Z.; Chen, B.; Wang, J.; Begovic, M.; Chen, C. Coordinated energy management of networked microgrids in distribution systems. *IEEE Trans. Smart Grid* **2015**, *6*, 45–53. [CrossRef]
36. Baran, M.E.; Wu, F.F. Network reconfiguration in distribution systems for loss reduction and load balancing. *IEEE Trans. Power Deliv.* **1989**, *4*, 1401–1407. [CrossRef]
37. Project 1090: Shangdong Changdao 27.2 MW Wind Power Project 2008. Available online: https://cdm.unfccc.int/Projects/DB/DNV-CUK1176964325.8/view (accessed on 21 January 2008).

38. Kim, H.Y.; Kim, M.K.; Kim, S. Multi-objective scheduling optimization based on a modified non-dominated sorting genetic algorithm-II in voltage source converter multi-terminal high voltage dc grid connected offshore wind farms with battery energy storage systems. *Energies* **2017**, *10*, 986. [CrossRef]
39. Mondai, D.; Chakrabarti, A.; Sengupta, A. Optimal placement and parameter setting of SVC and TCSC using PSO to mitigate small signal stability problem. *Int. J. Electr. Power Energy Syst.* **2012**, *42*, 334–340. [CrossRef]
40. Moghaddam, A.A.; Seifi, A.; Niknam, T.; Pahlavani, A.R.A. Multi-objective operation management of a renewable MG with back-up micro-turbine/fuel cell/battery hybrid power source. *Energy* **2011**, *36*, 6490–6507. [CrossRef]
41. Yuan, X.; Zhang, B.; Wang, P.; Liang, J.; Yuan, Y.; Huang, Y.; Lei, X. Multi-objective optimal power flow based on improved strength Pareto evolutionary algorithm. *Energy* **2017**, *122*, 70–82. [CrossRef]
42. Gerbex, S.; Cherkaoui, R.; Germond, A.J. Optimal location of multi-type FACTS devices in a power system by means of Genetic Algorithms. *IEEE Trans. Power Syst.* **2001**, *16*, 537–544. [CrossRef]
43. Marouani, I.; Guesmi, T.; Abdallah, H.H.; Ouali, A. Application of NSGA-II approach to optimal location of UPFC devices in electrical power systems. *J. Sci. Res.* **2011**, *10*, 592–603.

Article

The Effect of a Renewable Energy Certificate Incentive on Mitigating Wind Power Fluctuations: A Case Study of Jeju Island

Woong Ko [1], Jaeho Lee [1] and Jinho Kim [2,*]

[1]   Research Institute for Solar and Sustainable Energies, Gwangju Institute of Science and Technology, 123 Cheomdangwagi-ro, Buk-gu, Gwangju 61005, Korea; kwoong2001@gist.ac.kr (W.K.); jaeholee@buffalo.edu (J.L.)

[2]   School of Integrated Technology, Gwangju Institute of Science and Technology, 123 Cheomdangwagi-ro, Buk-gu, Gwangju 61005, Korea

*   Correspondence: jeikim@gist.ac.kr; Tel./Fax: +82-62-715-5322

Received: 11 March 2019; Accepted: 18 April 2019; Published: 20 April 2019

Abstract: As renewable energy penetration in power systems grows, adequate energy policies are needed to support the system's operations with flexible resources and to adopt more sustainable energies. A peak-biased incentive for energy storage systems (ESS) using the Korean renewable portfolio standard could make power system operations more difficult. For the first time in the research, this study evaluates the effect of imposing a renewable energy certificate incentive in off-peak periods on mitigating wind power fluctuations. We design a coordinated model of a wind farm with an ESS to model the behavior of wind farm operators. Optimization problems are formulated as mixed integer linear programming problems to test the implementation of revenue models under Korean policy. These models are designed to consider additional incentives for discharging the ESS during off-peak periods. The effects of imposing the incentives on wind power fluctuations are evaluated using the magnitude of the renewable energy certificate (REC) multiplier.

Keywords: renewable portfolio standard; renewable energy certificate; wind farm; energy storage system; variation criteria; mixed integer linear programming

---

1. Introduction

As global energy policies try to reduce the use of conventional fuel-based generators, the rate of renewable energy penetration has increased. Following global trends, South Korea has adopted a renewable portfolio standard (RPS) to promote the utilization of sustainable energies. In 2012, renewable promotion policies, such as feed-in tariffs (FITs), subsidized renewable energy owners, regardless of the actual power generation. The shift from FITs to the RPS was due to a financial shortage. In order to solve this problem, the RPS was designed to impose renewable energy generation obligations on large suppliers instead of offering unconditional energy subsidies [1]. Within this policy, renewable energy owners can retrieve their investment by obtaining a renewable energy certificate (REC) credit for the actual generation and selling it in the REC market.

Among many South Korean regions, Jeju Island has been aggressively adopting renewable energy as a way of achieving carbon-free power generation by 2030. With these efforts, renewable energy in Jeju exceeded 48.7% of the total energy generation in the winter of 2018 [2]. Although the increasing renewable energy supply is eco-friendly, the high fraction of energy produced by intermittent renewable energy sources, such as wind and solar, affects the power system's stability and reliability [3,4]. The increase in renewable energy use can cause a reduction in the power system's inertia that is necessary for reliable operation. This reduced inertia can easily result in large frequency fluctuations. Jeju, where electricity has been mostly supplied by high-voltage direct current (HVDC) connected to the mainland, is especially vulnerable to the variability in energy supply caused by wind power [5,6].

Wind farm operators should follow a grid code, defined by regulators, to connect their wind farm to the power grid safely. The grid code includes requirements for additional controller installation to control the voltage at the connected node and support the grid with a contingency energy reserve [7]. Among the terms of the code, power variation at the point of common coupling (PCC) due to wind power fluctuations is a major concern of wind farm operators and power system operators. European countries have set power gradient limits in their grid codes as the contribution from renewable energy generation increases [8]. Furthermore, though renewables are considered to be the cheapest energy sources, excess renewable energy generation has to be curtailed for operational stability [9]. The ramping that occurs at the PCC should also be considered since the reserve and ramping components are secured to prevent power shortages [10]. Although there are no requirements stipulated in the grid code for wind power fluctuations in Korea, it is expected to recommend that the power variation at the PCC in a 5 min period does not exceed 5% of wind farm capacity.

Among the methods to mitigate power fluctuations, utilizing an energy storage system (ESS) is considered to be a key solution for the power system operation of large wind farms. [11]. An ESS can handle the power variability with flexible charging and discharging. In many studies, performance analyses were conducted under intermittent power systems with various ESS types, including flywheels, superconductors, and fuel cells. Among energy storage sources, the battery-based ESS (BESS) demonstrated a better performance [12]; its advanced technology and large market share were favorable for large offshore wind farms [13].

With these advantages, the operation and the planning methods for an ESS have been introduced to improve wind farm operations. The coordinated control method for ESS integration with wind turbines was proposed to reduce wind power fluctuations and to extend ESS lifespan [14]. An adaptive supervisory control scheme was designed for wind turbine-integrated systems and ESSs to improve power qualities [15]. Besides designing a control scheme, optimal ESS scheduling strategies have been proposed to improve contract fulfillment and to minimize the curtailment of renewable energy production under uncertain circumstances [16,17]. For efficient wind farm operation, it is important to determine the ESS's size and utilize it efficiently. Methodologies for choosing an adequate storage size have been proposed to handle ramping events and mitigate scheduling errors caused by intermittent power generation [18]. Determining an installation site for an ESS is a major consideration in planning the coordinated system under the stochastic nature of power generation and load [19,20]. Furthermore, an optimal management policy was designed to maximize benefits and select an optimal storage size [21]. In short, the proper operation and planning strategies of an ESS can help wind farm operators and power system operators minimize their operational costs under an intermittent power supply.

Korea has encouraged ESS installation on wind farms through the provision of subsidies. According to the revenue agreements for wind power generation and ESS discharging power, revenue is dependent on peak and off-peak periods [22]. The power provided by the PCC is sold at a system marginal price (SMP), and the wind power generation transmitted to the PCC can earn an REC credit at all periods. However, during peak periods, wind farm operators can obtain additional REC credits for power discharged from their ESS. With this incentive policy, wind farm operators might recoup their ESS installation investment. In reality, however, peak-biased incentives affect power system operations since wind farm operators discharge their ESS power only during peak periods when additional revenue can be earned [23]. Although the power output of wind farms can fluctuate at all times, the lack of incentive for utilizing the ESS during off-peak periods may increase power variability as wind energy penetration increases. Hence, considering peak-biased incentive policies, ESSs cannot adequately support the power system at all periods.

While power system operators agree that utilizing energy storage resources is important for stable system operation, no study has focused on the effect of peak-biased incentives on wind power fluctuations. To address this, we evaluated the interaction between imposing an REC multiplier for discharging power during off-peak periods and the number of power shortages under Korea's renewable energy policy. This study addresses the following aspects:

- A coordinated model of a wind farm with an ESS is constructed to model actual behaviors of wind farm operators;
- A revenue optimization problem is modeled as a mixed integer linear programming (MILP) problem based on the settlement rules of wind power generation;
- An objective function is modified to consider additional revenue obtained by discharging in off-peak periods; and
- The effect of imposing an additional incentive in off-peak periods is analyzed according to the REC multiplier magnitude.

The rest of this paper is organized as follows: Section 2 describes an optimization model of a wind farm with an ESS. The modified optimization model to induce discharge in off-peak periods is introduced in Section 3. The effects of an REC multiplier in off-peak periods on mitigating wind power fluctuations are analyzed through a case study in Section 4. We conclude in Section 5.

2. An Optimization Model for Wind Farms with an ESS

2.1. Coordinated Model of Wind Farms with an ESS

This study examined a coordinated model of an onshore wind farm with an ESS. The farm was connected to an electricity grid via the PCC as shown in Figure 1.

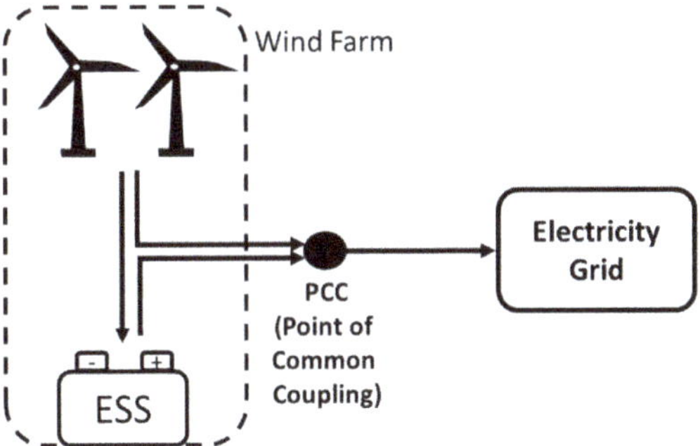

Figure 1. A coordinated model of a wind farm with an energy storage system (ESS).

Power generated from wind turbines can flow in two directions. One way is toward the ESS, where the power is used to charge the storage system. The rest of power is transmitted to the grid via the PCC. In other words, the power charge of the ESS comes only from wind power generation.

This model allows the wind farm operator to transmit and sell the generated electricity via the PCC to the main grid. The revenue from selling the electricity to the gird is comprised of two parts: first, is the income from wind power generation, excluding the power for battery-charging; and second, is the income from the ESS discharge.

This model assumes that the operator is not responsible for mitigating the resource variability since there are no penalties or incentives dependent on fluctuating wind power generation under the current policy.

2.2. Objective Function

Revenue and expense models were defined for describing the behaviors of the wind farm operator. The revenue models were composed of revenues in off-peak periods, R_t^{opk}, and peak periods, R_t^{pk}. In the expense model, C_t^{ESS}, could vary with the total amount of charged and discharged power. Hence, the objective function is given by:

$$\text{Maximize} \sum_{t \subset Off\ peak} R_t^{opk} + \sum_{t \subset peak} R_t^{pk} - \sum_{\forall t} C_t^{ESS} \ \forall t, \tag{1}$$

and, considering revenues in off-peak and peak periods and expense

$$R_t^{opk} = P_t^{PCC} \times smp_t + \left(P_t^{wind} - P_t^{ch}\right) \times REC_t \ \forall t, \tag{2}$$

$$R_t^{pk} = P_t^{PCC} \times smp_t + \left(P_t^{wind} - P_t^{ch}\right) \times REC_t + \alpha \times P_t^{dch} \times REC_t \ \forall t, \tag{3}$$

$$C_t^{ESS} = \left(P_t^{ch} + P_t^{dch}\right) \times \omega^{VO\&M} \ \forall t, \tag{4}$$

where the power at the PCC, P_t^{PCC}, is described as

$$P_t^{PCC} = P_t^{wind} - P_t^{ch} + P_t^{dch} \ \forall t. \tag{5}$$

The power at the PCC and the net power of the wind farm are paid out at the SMP and the REC price, respectively, regardless of the time period, as shown by the first and second variables of the right side of the Equations (2) and (3). The ESS discharged power during the peak period is paid out at the REC price by multiplying the REC multiplier, α, as depicted by the third variable from the right side of Equation (3). The variable costs of operating the ESS is composed of variable operation and maintenance costs and the total amount of charging and discharging as shown in Equation (5).

Using this optimization problem, wind farm operators can maximize revenue by controlling the amount of ESS charged and discharged power.

2.3. Operational Constraints of an ESS

The ESS can charge and discharge power within its state of charge (SOC) limit. Hence, the SOC of the ESS, SOC_t, is limited by the minimum and maximum SOC levels:

$$SOC^{\min} \leq SOC_t \leq SOC^{\max} \ \forall t. \tag{6}$$

The charging-discharging operations determine the SOC and the initial SOC is assumed to be half of the storage capacity:

$$SOC_t = \begin{cases} Cap^{ESS}/2 + P_t^{ch} \times \eta^{ESS,eff} - P_t^{dch}/\eta^{ESS,eff}, & if\ t = 1 \\ SOC_{t-1} + P_t^{ch} \times \eta^{ESS,eff} - P_t^{dch}/\eta^{ESS,eff}, & otherwise \end{cases}. \tag{7}$$

The ESS can only operate in charging-discharging mode at a single time-step as follows:

$$\delta_t^{ch} + \delta_t^{dch} \leq 1 \ \forall t. \tag{8}$$

The amount of discharging or charging power is limited by the capacity and the ESS charging-discharging efficiency. These characteristics are given by

$$0 \leq P_t^{ch} \leq \eta^{ch} \times \delta_t^{ch} \times Cap^{ESS}/5 \ \forall t \ \text{and} \tag{9}$$

$$0 \leq P_t^{dch} \leq \eta^{dch} \times \delta_t^{dch} \times Cap^{ESS}/5 \ \forall t. \tag{10}$$

The capacity multiplications and the ESS charging-discharging efficiency are divided by 5 since we assume that the ESS can be fully discharged or charged in 25 min [24].

The charging power can only be supplied from the wind farm when the following conditions are met:

$$0 \leq P_t^{ch} \leq P_t^{Wind} \ \forall t. \tag{11}$$

2.4. Variability Criteria for Wind Power Generation

The variability is defined as the power deviation at the PCC in a 5-min time step. The grid code model for the variability is described as follows:

$$-\sigma^{var} \times Cap^{Wind} \leq P_t^{wind} - P_{t-1}^{PCC} - P_t^{ch} + P_t^{dch} - P_t^{Excess} + P_t^{Short} \leq \sigma^{var} \times Cap^{Wind} \ t = 2,3,\cdots,t^{max}, \tag{12}$$

where σ^{var} is the wind farm capacity ratio of the variation criteria, and P_t^{Excess} and P_t^{Short} are the surplus variables for the upward and downward power, respectively, that exceed the variation criteria. The power at the PCC at time step t is defined by its representation in Equation (12) since the amount of ESS charging-discharging at time step t can change the power deviation of the adjacent time step. The surplus variables should have a non-zero value only if the power deviation at the two adjacent time steps exceeds the variability criteria in the opposite sides of the equation. Additional constraints for these surplus variables are required since Equation (12) is insufficient to determine the surplus. These constraints are described as follows:

$$P_t^{Excess} = \delta_t^{Excess}\left(P_t^{wind} - P_{t-1}^{PCC} - P_t^{ch} - \sigma^{var} \times Cap^{Wind}\right)t = 2,3,\cdots,t^{max} \ \text{and} \tag{13}$$

$$P_t^{Short} = \delta_t^{Short}\left(P_t^{wind} - P_{t-1}^{PCC} + P_t^{dch} + \sigma^{var} \times Cap^{Wind}\right)t = 2,3,\cdots,t^{max}, \tag{14}$$

where δ_t^{Excess} and δ_t^{Short} are binary variables representing the state of exceeding the upper and lower variation criteria, respectively, in time-step t. These variables are below 1 at a single time step, as follows:

$$\delta_t^{Excess} + \delta_t^{Short} \leq 1 \ t = 2,3,\cdots,t^{max}. \tag{15}$$

The surplus variables in Equations (13) and (14) mitigate the deviation between the wind farm generation at time step t and the power at the PCC at time-step $t-1$, along with the charging/discharging power at time-step t. In short, the surplus variables are assumed to be control variables to mitigate the variability in wind power generation.

The process of determining the value of the surplus variables is described in Figure 2, where the power deviation and the upper and lower criteria are observed.

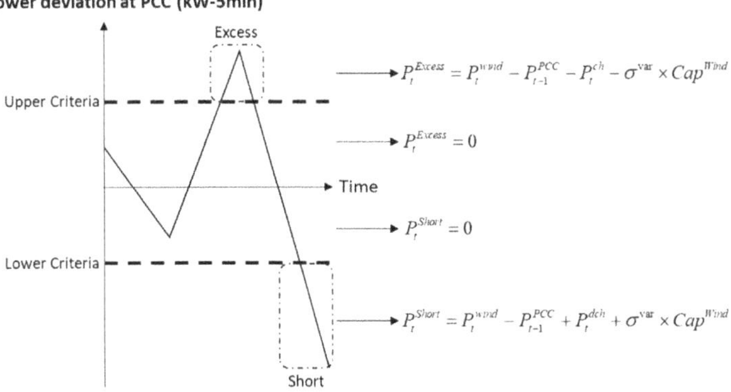

Figure 2. Determination of the surplus variable for power exceeding the upper and lower variation criteria.

The surplus variables can be determined by the upper and lower criteria. If the power deviation (the solid line) is within the criteria (the dotted line), these variables must be zero since the deviation does not exceed the criteria. In this case, the variables in Equations (13) and (14) are zero since both variables in Equation (15) are zero. However, if the deviation is outside of the criteria, the surplus variables should be the amount of power exceeding the criteria. In this case, the excess power is equal to variables P_t^{Excess} or P_t^{Short} in Equation (13) or (14), since the binary variable, δ_t^{Excess} or δ_t^{Short}, is 1.

In this optimization problem, Equations (13) and (14) reflect the above process and are depicted as nonlinear constraints. Therefore, MILP constraint linearization is required, as described in Appendix A.

3. A Modified Optimization Model to Induce the Discharge of an ESS in Off-Peak Periods

3.1. Modified Objective Function

The off-peak period revenue could be modified in order to induce the charging and discharging operations by the wind farm operator. Hence, the modified objective function is defined as:

$$Maximize \sum_{t \subset Off\ peak} R_t^{new\_opk} + \sum_{t \subset peak} R_t^{pk} - \sum_{\forall t} C_t^{ESS} \ \forall t, \tag{16}$$

and considering modified revenue

$$R_t^{new\_opk} = P_t^{PCC} \times smp_t + \left(P_t^{wind} - P_t^{ch}\right) \times REC_t + \alpha^{opk} \times P_t^{dch} \times REC_t \ \forall t, \tag{17}$$

where α^{opk} is the REC multiplier for discharging power in off-peak periods. The additional revenue generated is reflected in the third variable on the right side of the Equation (17). The peak period revenue and the expense of the ESS operation are the same as in Equations (3) and (4).

Conditions for determining α^{opk} are necessary to define the modified revenue and validate the profit for the operator.

3.2. Conditions for Determining REC Multiplier for Discharging Power in Off-Peak Periods

The REC multiplier for the discharging power in off-peak periods must be at least more than a specified value to initiate discharging. Therefore, it is important to consider the conditions for determining this multiplier as follows:

1. Condition 1: The REC multiplier value is more than 1; and
2. Condition 2: The REC multiplier value is more than the product reciprocal of the parameters related to ESS operations.

Condition 1 is the most common for determining the REC multiplier. As shown in Figure 1, wind farm operators can utilize their power resource generation in two ways. First, the generation transmitted to the electricity grid via the PCC can be paid out at a one-REC credit in the off-peak period, based on the settlement rules. In contrast, power generation for the ESS results in no profit for the operators. Charging loss occurs in ESS operations since the charging efficiency, which is less than 100%, governs the amount of charging power. In this case, the operators would sell their power generation to the main grid, rather than charge the battery, in order to increase revenue. Therefore, the REC multiplier value should be greater than 1 to be fully recognized as a whole power generation, regardless of charging loss.

Condition 2 is a supplemental condition for determining the REC multiplier, along with Condition 1. As shown in Equations (7), (9), and (10), the turnaround and discharging efficiency, as well as the charging efficiency, would result in a charging and discharging loss since these efficiencies are also less than 100%. First, the power loss would occur during the charging operation, with an efficiency of η^{ch}. The second loss is incurred by the inner operations of the ESS, modeled with the turnaround efficiency of $\eta^{ESS,\ eff}$. Discharging loss occurs when the ESS discharges its power, with an efficiency of η^{dch}. Although the power from the ESS and wind turbines both have the same effect of supplying electricity to the grid, the discharging power of the ESS is from partial wind power production with losses. The REC multiplier for the discharging power should be higher than that for the wind power to compensate for power losses. Hence, the minimum value for the REC multiplier in off-peak periods would be the reciprocal of the charging, turnaround, and discharging efficiencies.

4. Case Study and Discussion

The effectiveness of the proposed revenue model was evaluated by comparing the total discharge power and the total power exceeding the variation criteria with the conventional revenue model.

4.1. Simulation Setup

4.1.1. Parameters of a Wind Farm and an ESS

Operational data related to the coordinated model were required to simulate the coordinated wind farm model with an ESS as shown in Figure 1. The key parameters for the coordinated model operation are listed in Table 1.

Table 1. Key parameters of a wind farm and ESS operation.

Parameter	Value
Wind Farm Capacity [kW]	10,790
ESS Capacity [kWh]	2450
SOC^{min}/SOC^{max} [%]	20/80
Turnaround efficiency [%]	90
Charging/Discharging efficiency [%]	90/90
Variable O&M cost of ESS [Won/kW-5 min]	0.0275
REC multiplier for discharging power in the peak period	4.5

The wind farm in Jeju Island is a 10,790 kW Haengwon offshore farm [25]. The ESS in the wind farm was assumed to be installed with an optimized capacity of 2450 kWh [26]. The operation conditions for the ESS, regarding the minimum and maximum charge level, turnaround, and charging and discharging efficiency were assumed to be 90%. In addition, the variable O&M cost of ESS was adjusted to a five-minute time step unit [27]. The REC multiplier for the discharging power in peak periods followed the current RPS policy [22].

4.1.2. Wind Farm Generation Profiles

One-minute time step data for yearly wind power generation profiles were obtained from the Korea Power Exchange (KPX). The time step of these profiles was changed from one-minute to five-minute in order to match the optimization model time step. Additionally, characteristic days in these profiles were selected to decrease the computational burden. The selection procedure was as follows:

- The yearly data were sorted by month;
- Power exceeding the upper and lower variation criteria for daily wind power generation in the monthly profiles was calculated;
- The day with the highest number of occurrences beyond the variation criteria was chosen as a typical day in the monthly profile; and
- The above procedures were repeated for all months.

The selected dates, the number of violations of the variation criteria, the seasons, and months are listed in Table 2. The season was required to determine the REC multiplier during peak periods. In addition, the peak period for the REC multiplier is listed in the table.

Table 2. The number of violations of the variation criteria and peak period for the renewable energy certificate (REC) multiplier.

Season	Month	Date	Number of Violating Variation Criteria	Peak Period for REC Multiplier
Winter	1	01.20	34	
Winter	2	02.10	26	18:00~21:00
Winter	3	03.08	18	
Spring	4	04.17	15	
Spring	5	05.09	10	19:00~22:00
Spring	6	06.06	10	
Summer	7	07.02	6	
Summer	8	08.20	10	13:00~15:00, 19:00~21:00
Summer	9	09.07	9	
Autumn	10	10.01	18	18:00~21:00
Winter	11	11.24	17	18:00~21:00
Winter	12	12.05	32	

As shown in Table 2, the number of violations of the variation criteria in winter tended to be greater than in other seasons. It was expected that the wind in winter on Jeju Island was stronger than the winds in other seasons. As expected, the variation in wind power generation at each five-minute time step in winter tended to be greater than that in other seasons, as shown in Figure 3.

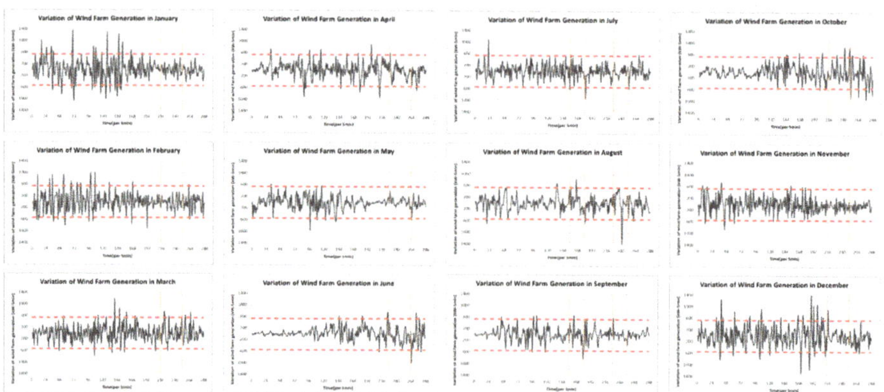

Figure 3. Profiles of wind power generation variation at each 5-min time step for each month based on one-year historical data at the Haengwon wind farm.

Figure 3 shows the variation profiles in rows from left to right in monthly order. The red dotted lines represent the variation criteria value. The gap between the yellow dotted lines signifies the peak period. Unlike other seasons, the wind power generation variation was unstable in winter from January to March and from November to December. The variation profiles in other seasons were mostly within the criteria. There were no noticeable trends in the variation profiles except in terms of seasons.

4.1.3. SMP and REC Price

The historical data for the yearly SMP and REC price was obtained for a one-hour time step [28,29]. Although the historical price differed both hourly and monthly, the average monthly price was calculated to reduce the variable price effect on the simulation, as shown in Table 3.

Table 3. Monthly average system marginal price (SMP) and REC price data.

Month	Average System Marginal Price [Won/kW-5 min]	Average REC Price [Won/kW-5 min]
1	8.85	20.51
2	10.6	17.72
3	11.06	17.49
4	10.11	17.28
5	10	19.07
6	10.39	17.38
7	10.12	17.92
8	9.39	17.88
9	9	19
10	8.87	16.44
11	10.79	18.05
12	10.77	14.59

The average REC price was nearly twice the average system marginal price. According to the table, the unit price was applied monthly in the simulation.

4.2. Case Study

4.2.1. Case Setup

The proposed MILP optimization problems were implemented in GAMS 25.1.2 and solved using GUROBI 8.0. The dual simplex algorithm terminated after reaching a 1% duality gap. This case study was conducted to compare the discharging power, the amount of excess power, and the total net revenue for the wind farm operator to determine whether to consider the REC multiplier in off-peak periods. When considering the REC multiplier, its values were listed by size, as shown in Table 4.

Table 4. Summary of the case setup.

Case Number	Status of REC Multiplier in Off-Peak Period	Magnitude of REC Multiplier
1	X	-
2	O	1
2-1	O	1.4
2-2	O	3
2-3	O	4.5

Case 1 is the reference case describing the current state of the wind farm operation without imposing the REC multiplier. Case 2 was the first control case where the REC multiplier was 1 for validating condition 1 as described in Section 3.2. The REC multiplier in case 2-1 was calculated by rounding up the reciprocal of charging, discharging, and turnaround efficiencies to two decimal places based on condition 2 as described in Section 3.2. Cases 2-2 and 2-3 were additional control cases for comparing the results with the increasing REC multiplier.

4.2.2. Case Results

From the power system operator's perspective, power shortages might be a major concern in operations since the shortage has to be supplemented by operating additional power resources. However, surplus power can be easily handled by curtailing wind power generation. Thus, the operator would pay attention to discharging patterns when power shortages occur. Therefore, this study focused on representing the results, including when the power exceeded the lower variation criteria and the discharging power.

The average discharging power in off-peak and peak periods was estimated to evaluate the effect of applying the REC multiplier, as depicted in Table 5.

Table 5. Average monthly discharging power in off-peak and peak periods.

Month	Average Discharging Power during Off-Peak Periods				Average Discharging Power during Peak Periods			
	Case 1 & Case 2	Case 2-1	Case 2-2	Case 2-3	Case 1 & Case 2	Case 2-1	Case 2-2	Case 2-3
1	0	163.8	158.5	160.5	182.4	182.9	172.8	130.2
2	0	156.7	157.4	129.4	184.5	187.2	137.2	71.5
3	0	160.6	161.1	161.3	182	179.9	155.5	161.7
4	0	110.4	153.3	151.1	181.3	181.3	172.5	152.5
5	0	155.6	162.2	166.8	179.1	179.3	179.3	146.4
6	0	138.7	155.8	155.8	183.1	183.9	178.7	152.6
7	0	158	152	147	183.5	198.9	175.2	169.2
8	0	154.4	157.5	157.7	186.7	190.3	186.3	180.6
9	0	151.9	152.6	149.9	192.1	191.3	192.1	183.1
10	0	147.8	148.5	151.8	178.4	180.4	178.2	153.6
11	0	121.7	158.7	161.4	181	182.6	163.5	148
12	0	163.8	155.1	163.2	177.3	178.8	164.8	142.3
Total Average	0	147.2	156.1	154.7	182.6	184.7	171.3	149.3

ESS operations under current policies could be modeled by case 1. As shown in Table 5, the ESS did not discharge in off-peak periods, while it discharged in peak periods. As expected, results of cases 1 and 2 are same. The results of the discharging power are first observed in case 2-1. In addition, the average discharging power during the peak period in case 2-1 was slightly higher than that in cases 1 and 2. This discharging power was presumed to increase due to altered ESS operations in the off-peak period. The average discharging power in the off-peak period in case 2-2 was higher than in case 2-1 due to an increased REC multiplier. However, the average discharging power in the peak period in case 2-2 was lower than that in case 2-1. These results are presumed to be due to the increase in discharging power during the off-peak period. Case 2-3 exhibited different trends. The average discharging power during the off-peak period in case 2-3 was slightly lower than that in Case 2-2. This result was compared with the increasing trend that existed in the discharging power during the off-peak period in cases 2-1 and 2-2. The average discharging power in the peak period in case 2-3 was lower than in the other cases.

The amount of power that exceeded the lower variation criteria is presented in Table 6. The effect of the REC multiplier on the power shortage is examined based on these case results.

Table 6. Monthly power exceeding the lower variation criteria of the point of common coupling (PCC) in off-peak and peak periods.

Month	Average Value of Power Exceeding the Lower Variation Criteria of PCC in the Off-Peak Period				Average Value of Power Exceeding the Lower Variation Criteria of PCC in the Peak Period			
	Case 1 & Case 2	Case 2-1	Case 2-2	Case 2-3	Case 1 & Case 2	Case 2-1	Case 2-2	Case 2-3
1	45.16	18.01	17.85	16.53	14.81	0	0	0
2	27.37	6.07	10.31	3.02	0	0	0	0
3	7.96	0	0	2.48	16.3	25.3	16.3	0
4	17.9	14.38	20.7	7.74	19.35	19.35	0	0
5	16.02	10.17	7.25	2.68	15.5	15.5	15.5	0
6	8.72	8.08	0	12.02	19.48	18.58	18.85	41.12
7	8.8	9.73	3.98	7.29	12.11	0	12.1	13.93
8	8.47	11.02	8.24	7.89	61.67	61.29	33.62	18.43
9	5	3.99	2.77	4.91	49.15	49.15	49.15	25.58
10	12.88	5.6	12.45	10.73	29.58	24.18	31.01	0
11	11.03	13.03	2.47	5.78	38.86	19.04	0	0
12	31.57	21.95	21.53	16.96	17.87	34.51	19.88	0
Total Average	16.74	10.17 (↓39.3%)	8.96 (↓46.5%)	8.17 (↓51.2%)	24.56	22.24 (↓9.4%)	16.37 (↓33.3%)	8.26 (↓66.4%)

In the off-peak period, ESS operations rarely affected the average power shortage in cases 1 and 2. The results of case 2-1, when the ESS begins discharging power, showed that the shortage was significantly reduced during the winter months. Furthermore, these monthly results demonstrated unique increments. Although there was a slight lack of consistency in the results, the total average power shortages decreased by almost 39% for cases 1 and 2. Similarly, in cases 2-2 and 2-3, the average value of power exceeding the lower variation criteria decreased, and the total average power for cases 1 and 2 reduced by almost 47% and 51%, respectively. The trends in power exceeding the variation criteria in the peak period are similar to those in the other period. Although the amount of discharging power seemed greater during the peak period, as shown in Table 5, the average power shortage in cases 1 and 2 were relatively larger in the peak period than those in the off-peak period. The shortages had a high value during the summer months since the peak period was longer than in other seasons. Contrary to the case trend of a decreasing discharging power during the peak period, the average value of power shortages decreased. In short, although the REC multiplier in the peak period was unchanged, the number of power shortages tended to decrease. These results imply that changes in the off-peak period would result in changes in the peak period by discharging power when the wind power fluctuations are significant.

The total monthly net revenue is presented in Table 7. Although the total net revenue results did not show a direct effect on mitigating the variation in wind power generation, the results suggest the effects of the additional incentive on the total net revenue.

Table 7. Total net revenue results.

Month	Case 1 & Case 2 [10^3 Won]	Case 2-1 [10^3 Won]	Case 2-2 [10^3 Won]	Case 2-3 [10^3 Won]
1	56,027	56,080	57,360	58,513
2	62,766	62,787	63,804	64,331
3	57,057	57,068	58,164	59,255
4	36,255	36,262	37,324	38,263
5	21,344	21,354	22,494	23,558
6	32,043	32,060	33,132	34,094
7	42,325	42,589	43,782	44,851
8	29,355	29,642	31,059	32,360
9	27,149	27,430	28,834	30,055
10	39,275	39,293	40,318	41,292
11	35,039	35,054	36,126	37,171
12	38,463	38,488	39,574	40,688

In Section 3.2, the results of Cases 1 and 2 are the same since the REC multiplier of 1 in off-peak periods could not induce discharging power. The revenue in case 2-1 increased slightly (0.01–1.02%). The REC multiplier applied in case 2-2 increased the revenue for cases 1 and 2 by 1.6–5.8%, and by 1.6–5.1% for case 2-1. Although the REC multiplier in case 2-2 was double that in case 2-1, a drastic revenue increase was not observed as the amount of the discharging power was assumed to be limited by the discharging rate. In case 2-3, the revenue was higher by 2.4–9.4% for cases 1, 2, and 2-1, and by 0.8–4.5% for case 2-2.

4.3. Discussions

The proposed revenue model for wind farms could reflect the current RPS policy in Korea as shown in results of case 1. From the other results, applying the REC multiplier in off-peak periods could be considered as a factor in reducing power shortages. This incentive would have a positive effect on mitigating wind power fluctuations. It might be also act as an economic strategy for power system operators since the additional revenue did not increase significantly, even when the magnitude of the REC multiplier was increased more than three-fold. Hence, the proposed incentive mechanism is a win-win proposition for wind farm and power system operators.

Further studies are required to analyze the operations and control strategies of ESSs since the proposed model was mainly designed for simulating revenue generation. The above results would be greatly influenced by the operation conditions of the ESS including SOC and charging and discharging rates. Although wind farm operators had the chance to gain more revenue as the REC multiplier increased, the additional revenue was limited since the amount of charging/discharging power was limited by these conditions. This limitation was considered as a factor in slight revenue increase despite a significant increase in the REC multiplier. Moreover, the proposed model could not show the optimal REC multiplier for both wind farm operators and power system operators. Therefore, future studies need to be designed to construct an optimization problem that reflects the detailed operations of the ESS and determines the optimal REC multiplier to control the power fluctuations.

5. Conclusions

This study evaluated the effect of an REC incentive on ESS operations to mitigate the variation in wind farm power generation on Jeju Island. We designed a coordinated model of the wind farm with an ESS to model the actual behaviors of a wind farm operator. From this model, an optimization problem was developed to maximize revenue for the operator. The objective function reflected the revenue from the SMP and the REC. The REC revenue function accounted for the discharging power, which paid out only in peak periods. Hence, the REC multiplier for discharging power in off-peak periods was applied to the modified REC revenue function. The conditions for determining the REC multiplier for the discharging power in off-peak periods were defined as they had not been previously accounted for. The problem constraints included the operational constraints of the ESS and the variability criteria for wind power generation. The linearization process was applied to the constraints of the variability criteria to be modeled as an MILP problem. Typical days in the one-year generation profiles were selected by using the variation criteria in order to ease the computational burden. Through the case studies, the modified revenue model induced ESS discharging in off-peak periods and reduced the average value of the power shortages without significant cost increases. We expect that applying the REC multiplier in off-peak periods would help wind farm and power system operators under intermittent renewable generation.

Author Contributions: All the authors contributed to this work. W.K. and J.L. designed the study and searched related researches. W.K. performed the analysis and wrote the first draft of the paper. J.K. contributed to the conceptual approach and thoroughly revised the paper.

Acknowledgments: This work was supported by the Korea Institute of Energy Technology Evaluation and Planning (KETEP) and the Ministry of Trade, Industry & Energy (MOTIE) of the Republic of Korea (No. 20181210301380).

Conflicts of Interest: The authors declare no conflict of interest.

Nomenclature

Indices and Sets

t	5-min time step in T.
T	Set of time steps.
$Off\,peak \subset T$	Subset of off-peak period.
$Peak \subset T$	Subset of peak period.

Parameters

smp_t	System marginal price at time, t, (Won/kW-5 min).
REC_t	Price of renewable energy certificate at time, t (Won/kW-5 min).
p_t^{Wind}	Wind farm power generation at time, t (kW-5 min).
Cap^{Wind}	Capacity of the wind farm (kW).
σ^{Var}	Variation criteria of the point of common coupling (PCC).
Cap^{ESS}	Capacity of the energy storage system (kWh).

$\omega^{VO\&M}$	Variable O&M cost of energy storage system (Won/kW-5 min).
SOC^{min}/SOC^{max}	Minimum/maximum state of charge level of the energy storage system (kWh).
$\eta^{ESS,eff}$	Turnaround efficiency of the energy storage system.
η^{ch}/η^{dch}	Charging/discharging efficiency of the energy storage system
α	Renewable energy certificate multiplier
α^{opk}	Renewable energy certificate multiplier in the off-peak period
Z	Positive infinity
Variables	
R_t^{opk}	Revenue in the off-peak period at time, t.
R_t^{pk}	Revenue in the peak period at time, t.
C_t^{ESS}	Operational cost of the energy storage system at time, t.
P_t^{PCC}	Power of the point of common coupling (PCC) at time, t.
P_t^{ch}/P_t^{dch}	Charging/discharging power of the energy storage system at time, t.
SOC_t	State of charge of the energy storage system at time, t.
P_t^{Excess}/P_t^{Short}	Surplus variable for the power exceeding the upper/lower variation criteria of the point of common coupling (PCC) at time, t.
Binary Variables	
$\delta_t^{ch}/\delta_t^{dch}$	Status of the charging/discharging operation of energy storage system at time, t.
$\delta_t^{Excess}/\delta_t^{Short}$	Status of exceeding the upper/lower variation criteria of the point of common coupling (PCC) at time, t.

Appendix A The Linearization Process of Nonlinear Constraints for Determining the Surplus Variable for Power Exceeding Upper–Lower PCC Variation Criteria

Equations (13) and (14) show the nonlinear constraints that determine the amount of power exceeding the upper and/or lower variation criteria. The right-side variables of these equations can be expressed as follows:

$$\begin{cases} P_t^{Excess} \geq P_t^{wind} - P_{t-1}^{PCC} - P_t^{ch} - \sigma^{var} \times Cap^{Wind} - \left(1 - \delta_t^{Excess}\right) \cdot Z \\ P_t^{Excess} \leq P_t^{wind} - P_{t-1}^{PCC} - P_t^{ch} - \sigma^{var} \times Cap^{Wind} + \left(1 - \delta_t^{Excess}\right) \cdot Z \end{cases} \quad t = 2, 3, \cdots, t^{max}, \tag{A1}$$

$$\begin{cases} P_t^{Excess} \geq -\delta_t^{Excess} \cdot Z \\ P_t^{Excess} \leq \delta_t^{Excess} \cdot Z \end{cases} \quad t = 2, 3, \cdots, t^{max}, \tag{A2}$$

$$\begin{cases} P_t^{Short} \geq P_t^{wind} - P_{t-1}^{PCC} + P_t^{dch} + \sigma^{var} \times Cap^{Wind} - \left(1 - \delta_t^{Short}\right) \cdot Z \\ P_t^{Short} \leq P_t^{wind} - P_{t-1}^{PCC} + P_t^{dch} + \sigma^{var} \times Cap^{Wind} + \left(1 - \delta_t^{Short}\right) \cdot Z \end{cases} \quad t = 2, 3, \cdots, t^{max}, \tag{A3}$$

$$\begin{cases} -P_t^{Short} \geq -\delta_t^{Short} \cdot Z \\ -P_t^{Short} \leq \delta_t^{Short} \cdot Z \end{cases} \quad t = 2, 3, \cdots, t^{max}. \tag{A4}$$

For the linearization process, either δ_t^{Excess} or δ_t^{Short} is equal to 0 if the power deviation at the PCC, expressed by $\left(P_t^{wind} - P_{t-1}^{PCC} - P_t^{ch}\right)$ or $\left(P_t^{wind} - P_{t-1}^{PCC} + P_t^{dch}\right)$, does not exceed the variation criteria, expressed by $\left(\sigma^{var} \cdot Cap^{Wind}\right)$. Then, the variable for the excessive power (P_t^{Excess} or P_t^{Short}) is 0, as shown in Equations (A2) and (A4). Otherwise, either δ_t^{Excess} or δ_t^{Short} is equal to 1. The power variable has a non-zero value as shown in Equations (A1) and (A3).

References

1. Jung, T.Y.; Kim, H.J. A critical review of the renewable portfolio standard in Korea. *Int. J. Energy Res.* **2016**, *40*, 572–578. [CrossRef]
2. Energy & Environment News. Renewable Energy Supplying Half Of Total Electricity Consumption in Jeju Island. Available online: http://www.e2news.com/news/articleView.html?idxno=204758 (accessed on 16 January 2019).
3. Jeju Energy Corporation. Q3 2018 Renewable Energy Plant Status. Available online: http://www.jejuenergy.or.kr/index.php/contents/open/development/development01?act=view&seq=1351&bd_bcid=development&page=1 (accessed on 17 January 2019).
4. Nguyen, N.; Mitra, J. Reliability of Power System with High Wind Penetration Under Frequency Stability Constraint. *IEEE Trans.Power Syst.* **2018**, *33*, 985–994. [CrossRef]

5. Yoon, M.; Yoon, Y.-T.; Jang, G. A Study on Maximum Wind Power Penetration Limit in Island Power System Considering High-Voltage Direct Current Interconnections. *Energies* **2015**, *8*, 14244–14259. [CrossRef]
6. Chang, B.; Ha, Y.; Jeon, W. Jeju Island System Planning Considering Wind Power Penetration with HVDC Links. *J. Int. Counc. Electr. Eng.* **2011**, *1*, 287–291. [CrossRef]
7. IRENA. Available online: https://www.irena.org/publications/2016/May/Scaling-up-Variable-Renewable-Power-The-Role-of-Grid-Codes (accessed on 22 January 2019).
8. Ikni, D.; Bagre, A.O.; Camara, M.B.; Dakyo, B. An offshore wind farm energy injection mastering using aerodynamic and kinetic control strategies. *E3S Web Conf.* **2018**, *61*, 00002. [CrossRef]
9. Collins, S.; Deane, P.; Ó Gallachóir, B.; Pfenninger, S.; Staffell, I. Impacts of Inter-annual Wind and Solar Variations on the European Power System. *Joule* **2018**, *2*, 2076–2090. [CrossRef] [PubMed]
10. California ISO. Flexible Ramping Product Uncertainty Calculation and Implementation Issues. Available online: https://www.caiso.com/Documents/FlexibleRampingProductUncertaintyCalculationImplementationIssues.pdf (accessed on 22 January 2019).
11. Zhao, H.; Wu, Q.; Hu, S.; Xu, H.; Rasmussen, C.N. Review of energy storage system for wind power integration support. *Appl. Energy* **2015**, *137*, 545–553. [CrossRef]
12. Mahto, T.; Mukherjee, V. Energy storage systms for mitigating the variability of isolated hybrid power system. *Renew. Sustain. Energy Rev.* **2015**, *51*, 1564–1577. [CrossRef]
13. Wang, X.; Li, L.; Palazoglu, A.; El-Farra, N.H.; Shah, N. Optimization and control of offshore wind systems with energy storage. *Energy Conv. Manag.* **2018**, *173*, 426–437. [CrossRef]
14. Kim, C.; Muljadi, E.; Chung, C.C. Coordinated Control of Wind Turbine and Energy Storage System for Reducing Wind Power Fluctuation. *Energies* **2018**, *11*, 52. [CrossRef]
15. Meghni, B.; Dib, D.; Azar, A.T.; Saadoun, A. Effective supervisory controller to extend optimal energy management in hybrid wind turbine under energy and reliability constraints. *Int. J. Dyn. Control* **2018**, *6*, 369–383. [CrossRef]
16. Damiano, A.; Gatto, G.; Marongiu, I.; Porru, M.; Serpi, A. Real-Time Control Strategy of Energy Storage Systems for Renewable Energy Sources Exploitation. *IEEE Trans. Sustain. Energy* **2014**, *5*, 567–576. [CrossRef]
17. Choi, S.; Min, S. Optimal Scheduling and Operation of the ESS for Prosumer Market Environment in Grid-Connected Industrial Complex. *IEEE Trans. Ind. Appl.* **2018**, *54*, 1949–1957. [CrossRef]
18. Hartmann, B.; Dán, A. Methodologies for Storage Size Determination for the Integration of Wind Power. *IEEE Trans. Sust. Energy* **2014**, *5*, 182–189. [CrossRef]
19. Nick, M.; Cherkaoui, R.; Paolone, M. Optimal siting and sizing of distributed energy storage systems via alternating direction method of multipliers. *Int. J. Electr. Power Energy Syst.* **2015**, *72*, 33–39. [CrossRef]
20. Delgado-Antillón, C.P.; Domínguez-Navarro, J.A. Probabilistic siting and sizing of energy storage systems in distribution power systems based on the islanding feature. *Electr. Power Syst. Res.* **2018**, *155*, 225–235. [CrossRef]
21. Choi, D.G.; Min, D.; Ryu, J.-H. Economic Value Assessment and Optimal Sizing of an Energy Storage System in a Grid-Connected Wind Farm. *Energies* **2018**, *11*, 591. [CrossRef]
22. Korea Energy Agency. Renewable Portfolio Standards(RPS). Available online: http://www.energy.or.kr (accessed on 13 December 2018).
23. Energy & Environmental News. Dilemma on Integrating the ESS with the Wind Farm in Jeju Island. Available online: http://www.e2news.com/news/articleView.html?idxno=94724 (accessed on 22 January 2019).
24. Samsung SDI. ENERGY STORAGE SYSTEM for Utility, Commercial. Available online: http://www.samsungsdi.co.kr/upload/ess_brochure/ESS%20for%20Utility%20Commercial.pdf (accessed on 14 December 2018).
25. Jeju Energy Corporation. Wind Facilities. Available online: http://www.jejuenergy.or.kr/index.php/contents/energy/facilities (accessed on 14 December 2018).
26. Kang, M.-S.; Jin, K.-M.; Kim, E.-H.; Oh, S.-B.; Lee, J.-M. A Study on the Determining ESS Capacity for Stabilizing Power Output of Haeng-won Wind Farm in Jeju. *J. Korean Solar Energy Soc.* **2012**, *32*, 25–31. [CrossRef]

27. Platte River Power Authority. Battery Energy Storage Technology Assessment. Available online: https://www.prpa.org/wp-content/uploads/2017/10/HDR-Battery-Energy-Storage-Assessment.pdf (accessed on 14 December 2018).
28. Korea Power Exchange. REC Price. Available online: http://onerec.kmos.kr/portal/rec/selectRecReport_tradePerformanceList.do?key=1971 (accessed on 14 December 2018).
29. Korea Power Exchange. SMP for Jeju. Available online: http://www.kpx.or.kr/www/contents.do?key=226 (accessed on 14 December 2018).

Article

An Optimized Protection Coordination Scheme for the Optimal Coordination of Overcurrent Relays Using a Nature-Inspired Root Tree Algorithm

Abdul Wadood [1], Saeid Gholami Farkoush [1], Tahir Khurshaid [1], Chang-Hwan Kim [1], Jiangtao Yu [1], Zong Woo Geem [2] and Sang-Bong Rhee [1,*]

[1] Department of Electrical Engineering, Yeungnam University, 280 Daehak-Ro, Gyeongsan, Gyeongsangbuk-do 38541, Korea; wadood@ynu.ac.kr (A.W.); saeid_gholami@ynu.ac.kr (S.G.F.); tahir@ynu.ac.kr (T.K.); kranz@ynu.ac.kr (C.-H.K.); yujiangtao0221@gmail.com (J.Y.)

[2] Department of Energy IT, Gachon University, Seongnam 461-701, Korea; zwgeem@gmail.com

* Correspondence: rrsd@yu.ac.kr; Tel.: +82-10-3564-0970

Received: 13 August 2018; Accepted: 13 September 2018; Published: 15 September 2018

Abstract: In electrical engineering problems, bio- and nature-inspired optimization techniques are valuable ways to minimize or maximize an objective function. We use the root tree algorithm (RTO), inspired by the random movement of roots, to search for the global optimum, in order to best solve the problem of overcurrent relays (OCRs). It is a complex and highly linear constrained optimization problem. In this problem, we have one type of design variable, time multiplier settings (TMSs), for each relay in the circuit. The objective function is to minimize the total operating time of all the primary relays to avoid excessive interruptions. In this paper, three case studies have been considered. From the simulation results, it has been observed that the RTO with certain parameter settings operates better compared to the other up-to-date algorithms.

Keywords: nature-inspired optimization; root tree algorithm (RTO); time multiplier setting (TMS); overcurrent relay (OCR); protection scheme

---

1. Introduction

The power system consists of three main parts: generation, transmission, and distribution of electrical energy and works at voltage levels ranging from 415 V to 400 KV or greater. Moreover, the supply lines, which transmit electrical power, are not insulated. These lines show irregularities more often than other parts of the power system due to several issues, such as lightning, which lead to overcurrents. The imbalance due to these irregularities interrupts the behavior of power and, moreover, leads to the impairment of the other accessories linked to the power system. In order to eliminate these problems, protective measures should be taken into account. To improve this difficulty, overcurrent relays (OCRs) have been commonly used for years as a safety measure in the power system to avoid mal-operation in the power supply. Thus, OCRs are easy to use for the safety of the sub-transmission power system and a secondary layer of backup protection in transmission systems. During the designing of the electrical power system, it is important to consider the coordination of these OCRs. In case any fault occurs in the power system, OCRs help the breaker to play the logical elements and these relays are placed at both ends of the line. The OCRs decide which part of the power system has to come into action during a fault so that the faulty section is isolated and does not interrupt the power system, with certain constraints like adequate coordination tolerance and without extra disruptions. This procedure is mainly based on the nature of relays and other protection measures. The design variable of time dial settings (TDSs) for each relay in the circuit needs to be optimized. After optimizing the TDS, the faulty lines in the power system are isolated, thus ensuring a continuous supply of power to the remaining parts of the system. In the beginning, a trial-and-error methodology was utilized by scholars, which utilized a huge number of repetitions to reach an optimum relay setting. Thus, many researchers and scholars implemented the setting of OCRs based on experience. In [1], the authors applied the linear programming algorithm to tackle the optimum coordination of OCR. In this paper the authors changed the nonlinear issue into a linear problem with the assistance of factors without any estimation. The problem is solved linearly to determine the TDS and plug setting (PS) factors. After the linear solution is achieved, the roots of these nonlinear expressions (TDS and PS) must be found to achieve the solution. In [1], the utilization of optimization techniques was first proposed for this issue. A literature review on this problem can be found in [2] where a random search algorithm (RST-2) was connected to handle the issue of directional overcurrent relays (DOCRs), utilizing distinctive single- and multi-circle circulation relay models, individually. Additionally, the issue of DOCRs was handled with various techniques. For example, in [3], a protection scheme was designed based on directional current protection using the inverse time characteristic. In [4], a protection model was designed considering different modes of generation for the distributed generator. In [5], a microprocessor-based relay is investigated for the protection of micro-grids. In [6], a sparse dual revised simplex algorithm was used to optimize the TDS settings. In [7], a stable operation of distributed generation was conducted with help of optimal coordination of double-inverse overcurrent relay. The transient constrained protection coordination stagey has been used for a distribution system with distributed generation in [8]. In [9], a clustering topology has been used to reduce the number of different setting for adaptive relays coordination. In [10–13], other techniques of linear programming were used to solve the problem of DOCRs to optimize the TDS and PS settings, such as the simplex algorithm and the Rosen Brock hill-climbing algorithm.

Bio-inspired algorithms (BIAs) and artificial intelligence (AI)-based algorithms have been gaining the interest of scholars. In [14], a firefly algorithm was utilized to solve the problem of DOCR by taking some case studies of the IEEE standard bus system. The BIAs, which have been utilized to handle the issue of DOCRs, incorporate, yet are not constrained to, particle swarm optimization (PSO) [15,16], In [17], a genetic algorithm (GA) was introduced to solve the DOCR coordination problem. In [18], a grey wolf optimizer was used to find the accurate coordination between relays. In [19], modified evolutionary programming and evolutionary programming have been used for relay coordination. In [20], a modified electromagnetic field optimization algorithm has been used to solve the problem of DOCRs. In [21], an improved search algorithm was used to solve the relay coordination problem. Expert systems [22–25] and fuzzy logic [26] use AI calculations to handle the issue of DOCRs. The nature-inspired algorithms are not only restricted to overcurrent relay coordination but also have significant applications in different fields. Due to this, it has attracted the attention of scholars [27–31]. The significant drawback of the early proposed algorithm, including both the numerical and metaheuristic methodologies, is the probability of converging to values which may not be a global optimum but, rather, are stuck at a local optimum. To unravel this issue, an RTO is inspected in this examination for the ideal coordination of OCRs and is contrasted with chaotic firefly algorithms (CFA), firefly algorithms (FA), continuous particle swarm optimization (CPSO), continuous genetic algorithms (CGA), genetic algorithms (GA), and the simplex method (SM). More recently, the authors applied the root tree algorithm (RTO) to the DOCRs and compared them with different metaheuristic algorithms [32]. In this article, the problem of OCRs in a power system is handled by the RTO algorithm. The objective of this problem is to minimize the total operating time taken by primary relays. Primary relays are expected to isolate the faulty lines satisfying the constraints on design variables.

RTO is a renowned and trustworthy bio-inspired algorithm for solving linear, nonlinear, and complex constrained optimization problems. To the extent of the author's knowledge, RTO has not yet been executed for the advancement of OCR settings utilizing single- and multi-loop distribution bus systems, which are introduced in this paper. We have actualized RTO to take care of the issue of OCR settings, and the results of our simulation are contrasted with other up-to-date algorithms.

The rest of the paper is organized as follows. In Section 2, we elaborate on the mathematical formulation of the problem. The RTO is recalled in Section 3. Parameter settings and statistical and graphical results are discussed in Section 4. Section 5 concludes the present study.

2. Problem Formulation of the Overcurrent Relay

The coordination of the overcurrent relay is defined as an optimization issue in a multi- or single-source loop system. Nonetheless, the coordination issue has an objective function and limitations that should fulfill the distinct constraints:

$$\min f = \sum_{j=1}^{n} w_j T_{j,k} \tag{1}$$

where the parameters w_j and $T_{j,k}$ are the weight and operating time of the relay, respectively. For all the relays, $w_j = 1$ [33]. Therefore, the characteristic curve for operating relay R_i can be chosen from a portion of the selectable decision of IEC norms and could be characterized as:

$$T_{op} = TMS_i \left(\frac{\alpha}{\left(\frac{If_j}{Ip_j} \right)^k - 1} \right) \tag{2}$$

where α and k are steady parameters which characterize the relay characteristics and are expected to be $\alpha = 0.14$ and $k = 0.02$ for normal inverse type relay. The factors TMS_i and Ip_j are the time multiplier setting and pickup current of the *ith* relay, respectively, while If_i is the fault current flowing through relay R_i.

$$PSM = \frac{I_{fj}}{Ip_j} \tag{3}$$

where PSM remains for the plug setting multiplier and Ip_j is the primary or main pickup current.

$$T_{op} = TMS_i \left(\frac{\alpha}{(PSM)^k - 1} \right) \tag{4}$$

The above issue mentioned in condition (4) is a nonlinear issue in nature. The coordination can be planned as linear programming by considering the plug setting of the relay and the working time of the relays, a linear function of *TMS*. In linear programming, the *TMS* is ceaseless while rest of the parameters are steady, so condition (4) moves toward becoming:

$$T_{op} = a_\rho(TMS_i) \tag{5}$$

where:

$$a_\rho = \frac{\alpha}{(PSM)^k - 1} \tag{6}$$

Hence, the objective function can be formulated as:

$$\min f = \sum_{i=1}^{n} a_\rho(TMS_i) \tag{7}$$

Constraints

The total operating time could be minimized under two kinds of constraints, including the constraints of the relay parameter and constraints of coordination. The first constraints contain the limits of TMS while the other constraints are appurtenant to the coordination of the main and secondary relays. The limit on the relay setting parameters necessitates constraints in light of the decisions of relay parameter and design, and the points of confinement can be communicated as follows:

$$TMS_i{}^{min} \leq TMS_i \leq TMS_i{}^{max} \tag{8}$$

However, for a reasonable security margin, the pickup current should be lower than the short-circuit current and should be greater than the highest load current simultaneously. The other constraints are convenient for the acclimation of primary and secondary relay operating time. The coordination should be kept in a proper way where the secondary operating time of the relay should be greater than that of the primary relay. If the primary relay fails to clear the fault, the secondary relay should initiate its operation by opening the circuit breaker within a certain interval known as the coordination time interval (CTI). The coordination topology is shown in Figure 1. When a fault F takes place, it is sensed by both relays R_i and R_j. The primary relay responds first, having less operating time than that of the backup relay.

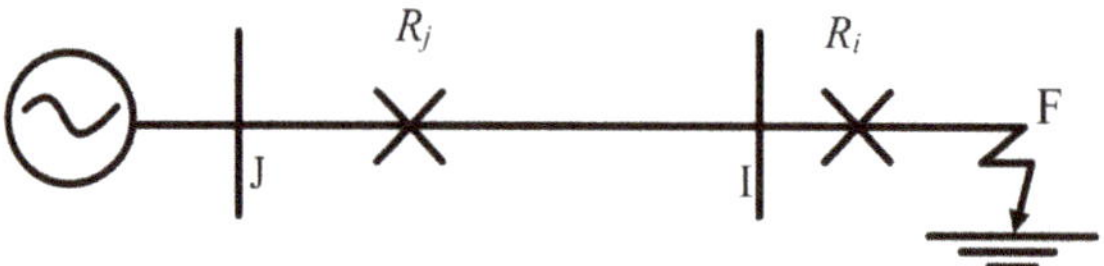

Figure 1. A single-end radial distribution system.

For fault F, beyond bus bar *i*, relay R_i should respond first to clear the fault. The operating time of R_i is set to 0.1 s. The backup relay should wait for 0.1 s plus the operating time of the circuit breaker at bus bar *i* and overshoot time of the relay R_j [34]. To maintain selectivity of primary and backup protection and to keep the grading of relay coordination accurate, the operating time of the backup relay (R_j) should always be greater than that of primary relay (R_i) by an amount in which R_i was supposed to have cleared the fault. This time is usually the sum of operating time of R_i and the breaker operating time. The current and voltage wave for the primary/back up pair are shown in Figure 2 for the typical single-end ring main feeder used in the distribution system.

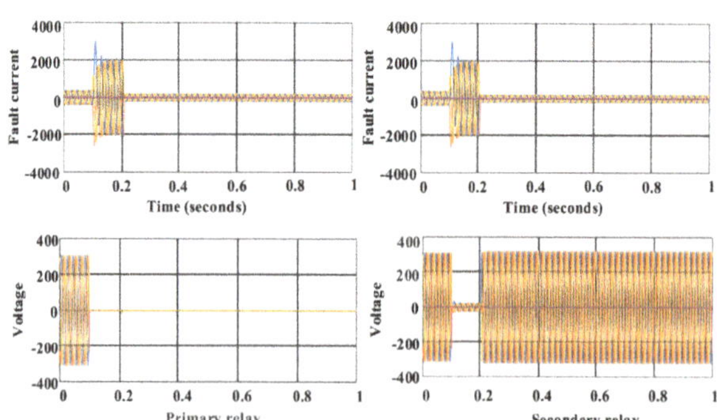

Figure 2. Voltage and current waveform of the primary/backup pair of the relay for the single-end radial system.

Hence, the coordination constraints could then be defined as:

$$T_j \geq T_i + CTI \tag{9}$$

where the parameter T_i and T_j are primary and backup relay operating times, respectively.

3. Root Tree Optimization Algorithm

The rooted tree optimization technique is a natural and biological algorithm inspired by the random-oriented movement of roots developed by Labbi Yacine et al. in 2016 [35]. Roots work in a group to find the global optima and the best place to get water, instead of working individually. However, an individual root has a limited capacity and a greater number of roots are based around the place which links the plant with the source of water. To design the algorithm, an imaginary nature of roots should be taken into consideration for their combined decision related to the wetness degree, where the head of the root is located. To find one or more wetness locations by random movement, these roots call other roots to strengthen their presence around that position to become a new starting point for majority of root groups to get to the original place of water that will be the optimal solution. Figure 3 shows the behavior of the roots of a plant that how they search for water to find the best location. The roots which are far or have less wetness degree are replaced by new roots which are oriented randomly. However, the roots which have a greater amount of wetness ratio will preserve their orientation, whereas roots (i.e., solutions) that are far from the water location could be replaced by roots near the best roots of the previous generation.

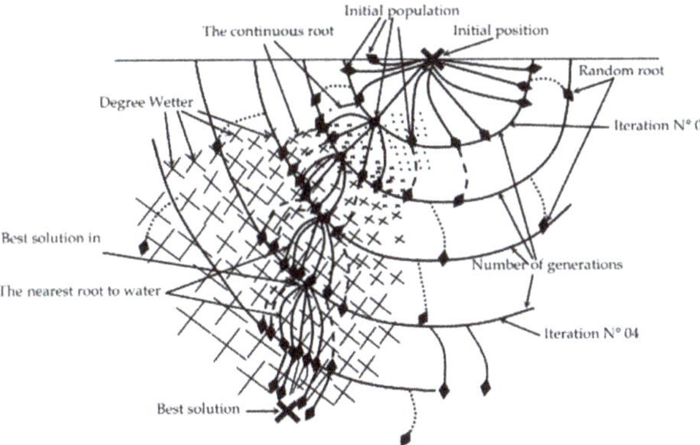

Figure 3. Searching mechanism of the RTO algorithm for searching for the water location [35].

However, the completion of a maximum number of cycles or generations is the stopping criterion at the end of generations. The solution with best fitness value will be the desired solution. The evaluation of population members is based on a given objective function which is assigned with its fitness value. The candidate with the best solutions is forwarded to next generation while the others are neglected and reimbursed by new group of random solutions in each iteration/generation. The proposed algorithm starts its work by creating an initial population randomly. However, for the RTO algorithm, there are some important parameters that need to be defined regarding how roots start the random movement from the initial population to new population. Those parameters are roots and wetness degree (wd). These two parameters can give the suggested solution and fitness value to the rest of the population. In this work, the RTO algorithm is used to find the optimal solution for the relay coordination problem within a power system. The suggested procedure has an extraordinary exploration ability and merging promptness in contrast with other methodology. This distinguishing property makes the populace individual from RTO more discriminative in locating the ideal result compared to that of another developmental method. The primary point of this paper is to locate the ideal estimations of TMS, keeping in mind the end goal to limit the aggregate working time of OCRs under a few requirements like relay settings and backup constraints.

3.1. The Rate of the Root (R_n) Nearest to the Water

The term rate here represents those roots which are a member of the total population that gather around the wetter place. This root will be next in line to that root where wetness is weaker than that of the previous generation. The fresh population of the nearest root according to wetness condition can be defined as:

$$x''(i, it + 1) = x_{it}^b + \frac{(k_1 * w_{di} * randn * l)}{n * it} \tag{10}$$

where it is the current step of the iteration, $x''(i, it + 1)$ is the fresh member for the next iteration, i.e., $(it + 1)$. The best solution achieved from the previous generation can be represented by x_{it}^b, while the parameters k_1, N, i, and l represent the adjustable parameter, population scale number, and upper limit, respectively, and $randn$ is a normal random number having a value between -1 and 1.

3.2. The Rate of the Continuous Root (R_c) in Its District

It is the root which has greater wetness ratio and moves forward from the previous generation. The new population of the random root is expressed as follows:

$$x''(i, it + 1) = x_{it}^b + \frac{(k_2 * w_{di} * randn * l)}{n * it} * \left(x^b(it) - x(i, it) \right) \tag{11}$$

whereas k_2 is the adjustable parameter while $x(it)$ is the iteration for the previous candidate at iteration it and $randn$ is a random number that has a range between 0 and 1.

3.3. Random Root Rate (R_r)

In this case, the roots spread randomly to find the best location of water to achieve the maximum rate of getting the global solution. The roots with less ratio of wetness degree are also replaced from the last generation. In this step, a new population calculated for a random root could be expressed as:

$$x''(i, it + 1) = x_r it + \frac{(k_3 * w_{di} * randn * l)}{it} \tag{12}$$

where the parameter x_r is the candidate which is selected randomly from the previous generation with an adjustable parameter k_3.

3.4. The Step Root Tree Algorithm

The steps of this algorithm can be compiled as follows:

Step 1: In this step, the initial generation is created randomly and is composed of N members with the variable limits in the research location with induce numerical values of R_n, R_r, and R_c rates.

Step 2: In the second step, all population members are measured based on their respective wd for the maximum and minimum objective function:

$$w_{di} = \begin{cases} \frac{f_i}{\max(f_i)} & \text{For the maximum objective} \\ 1 - \frac{f_i}{\max(f_i)} & \text{For the minimum objective} \end{cases} \quad i = 1, 2, \ldots, N \tag{13}$$

Step 3: The new population is created and replacement of the member is done according to the wetness ratio of R_n, R_r, and R_c. Equations (10)–(12) are used for a candidate having the smallest value until the one with the same wetness ratio is achieved.

Step 4: In this step, the meeting criteria are satisfied to display an optimal result with the best fitness value. Otherwise there is a return to Step 2.

3.5. Implementation of the RTO Algorithm for the OCR Coordination Problem

Step 1: Define the input parameter that includes TMS, the maximum number of iterations, population size, and the adjustable parameter (i.e., different rate values of R_n, R_r, and R_c).

Step 2: In this step, the population is initially generated with respect to the relay bounding. These individuals must be a feasible candidate solution that satisfies the relay operating constraints.

Step 3: Measure the fitness value of each candidate x_b in the population, with the help of Equation (10). Additionally, an evaluation of wetness ratio is done in this step for each candidate.

Step 4: In this step, the comparison is done between the best fitness values of candidates $\left(x_b{}^{best}\right)$ of the individual population.

Step 5: Computation of new candidates can be done using Equations (10)–(12).

$$X_b(i+1) = (X_{bi,r} + w_{di} + k_3)\frac{randn\left(X_{bi,max}-X_{bi,min}\right)}{it} \text{ for } i = 1, \ldots, R_r \times n$$

$$X_b(i+1) = \left(\left(X^{best}{}_b(i) + w_{di} + k_1\right)\right)\frac{rand\left(X_{bi,max}-X_{bi,min}\right)}{n*it} \text{ for } i = (R_r * n) + 1, \ldots, n \times (R_r + R_n)$$

$$X_b(i+1) = ((X_b(i) + w_{di} + k_2))randn\left(X^{best}{}_b - X_{bi,min}\right) \text{ for } i = (R_n + R_c) \times n + 1, \ldots, n \qquad (14)$$

where $X_{bi,r}$ is the individual that is randomly selected from the current population, n is the population size, $X_{bi,max}$, and $X_{bi,max}$ are upper and lower parameter limits, respectively, and i is the number of iterations.

Step 6: If the number of iterations reaches the maximum, then go to Step 7. Otherwise, go to Step 3.

Step 7: The candidates that generate the latest $x_b{}^{best}$ is the optimal TMS of each relay with the minimum total operating time that is the objective function.

The implementation and pseudocode of RTO for solving the coordination problem of OCR is depicted in Figure 4 and Algorithm 1.

Algorithm 1. The structure of the new rooted tree algorithm [35].

```
Algorithm RTO
Begin
// Initialization:
Set the rates Rn, Rr and Rc parameters
Give the maximum number of iterations: MaxIte, and the population scale: the RTO size
Set iteration counter it = 1
Generate the initial population X(1) randomly within the search range of (Xmin, Xmax)
// Loop
Repeat
Evaluate the wdi for each root // wd: Fitness, root: Individual
Reorder the population according to the witness degree
Identify the candidate according to the wetness place
Xbest // the global best in the whole population
For i = 1 to Rr the RTO size do
Selected individual Xr it randomly from the current population
Xi (it + 1) = Xr (it) + k1* wdi * randn * | Xmax + Xmin | /it
End for
For i= Rr * the RTO size +1 to (Rr+Rn) the RTO size do
Xi(it+1) = Xbest + k3* wdi *randn * | Xmax + Xmin | /(it theRTOsize)
End for
For i = (1 − Rc) the RTO size +1 to the RTO size do
Xi (it+1) = Xi (it) +k1* wdi *rand * (Xbest − Xi (it))
End for
Update Xbest
it = it + 1
Until for a stop criterion is not satisfied // & it <MaxIte
End
```

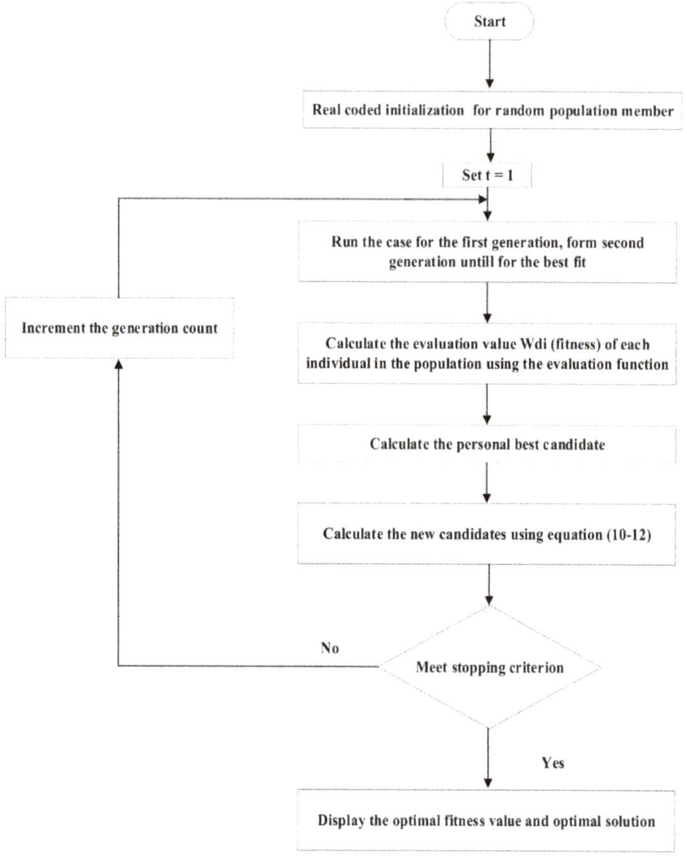

Figure 4. The flow chart of the RTO algorithm.

4. Results and Discussion

In this segment, a legitimate program has been created in MATLAB software to locate the optimum value of OCR in a single- and multi-loop distribution system utilizing RTO. The productivity and execution of RTO were tried for the distinctive single- and multi-loop system, and it was discovered that the RTO gives the most tasteful and best solution in all the contextual analyses. Three contextual analyses have been viewed. The system details of all contextual investigations could be found in [36–39]. In each case study, the following RTO parameters were used. The comprehensive explanation of the problem formulation and application of RTO to find the optimal solution is presented for all case studies. For the purpose of this study, population size = 50 and the maximum number of iteration = 200.

4.1. Case Study 1

Figure 5 shows a multi-loop system having eight overcurrent relays. The configuration and combustion of the primary and secondary relay pair models depend upon the location of the fault current in different feeders. In this combination, six different fault locations are taken into consideration. The aggregate fault current and the essential reinforcement relationship of the relay for the six fault focuses are given in Table 1. All relays have a plug setting of 1 and a CT ratio of 100:1. Table 2 shows the current seen by relays and the a_ρ constant for various faults.

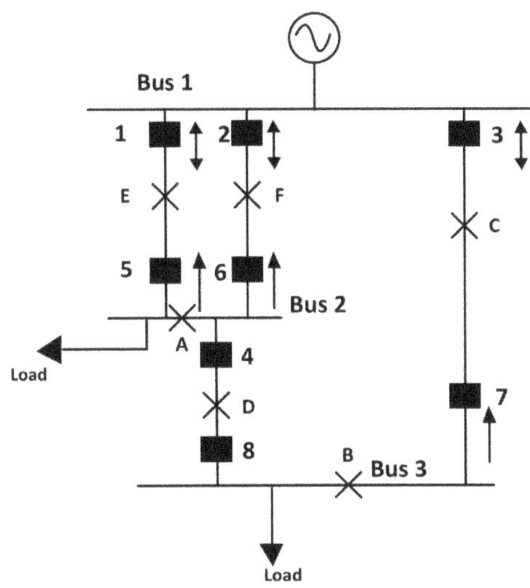

Figure 5. A multi-loop distribution system.

Table 1. Total fault current and primary/backup relationships of relays for Case 1.

Fault Point	T. Fault Current	Primary Relay	Backup Relay
A	2330	1, 2, 8	-, -, 3
B	1200	3, 4	-, 1, 2
C	1400	3, 7	-, 4
D	1400	4, 8	1, 2, 3
E	2800	1, 5	-, 8
F	2800	2, 6	-, 8

- Indicates no backup relay.

Table 2. a_ρ constants and relay currents for Case 1.

Fault Point		Relay							
		1	2	3	4	5	6	7	8
A	I_{relay}	10	10	3.3	-	-	-	-	3.3
	a_ρ	2.971	2.971	5.749	-	-	-		5.749
B	I_{relay}	3.45	3.45	5.1	6.9	-	-	-	-
	a_ρ	5.584	5.584	4.227	3.551	-	-	-	-
C	I_{relay}	2	2	10	4	-	-	4	-
	a_ρ	10.035	10.035	2.971	4.9804	-	-	4.9804	-
D	I_{relay}	5	5	4	10	-	-	-	4
	a_ρ	4.281	4.281	4.9804	2.971	-	-	-	4.9804
E	I_{relay}	20	6	2	-	8	-	-	2
	a_ρ	2.267	3.837	10.035	-	3.297	-	-	10.035
F	I_{relay}	6	20	2	-	-	8	-	2
	a_ρ	3.837	2.267	10.035	-	-	3.297	-	10.035

4.1.1. Mathematical Modelling of the Problem Formulation

In this case, eight variables had been taken into consideration (i.e., the TMS of eight relays), eight constraints due to bounds on relay operating time (or bounds on the TMS of relays), and nine constraints due to coordination criteria. The duplex of the proposed problem will have 17 variables and 8 constraints. In order to validate the efficiency of RTO, the CTI ratio of 0.6 s and minimum operating time (MOT) of the relay was taken as 0.1 s. The TMS of all the eight relays are $x1$–$x8$.

The problem for optimization can be demonstrated as:

$$minz = 28.975x_1 + 28.975x_2 + 37.736x_3 + 11.502x_4 + 3.297x_5 + 3.297x_6 + 4.9807x_7 + 30.7994x_8 \quad (15)$$

The constraints owing to the MOT of the relays are:

$$2.971x_1 \geq 0.1 \quad (16)$$

$$2.971x_2 \geq 0.1 \quad (17)$$

$$5.584x_3 \geq 0.1 \quad (18)$$

$$4.980x_4 \geq 0.1 \quad (19)$$

$$3.297x_5 \geq 0.1 \quad (20)$$

$$3.297x_6 \geq 0.1 \quad (21)$$

$$4.980x_7 \geq 0.1 \quad (22)$$

$$10.035x_8 \geq 0.1 \quad (23)$$

Hence, the constraints mentioned in Equations (18), (19), (22), and (23) violate the constraints of the minimum value of the (TMS). Hence, these constraints are reconstructed as:

$$x_3 \geq 0.025 \quad (24)$$

$$x_4 \geq 0.025 \quad (25)$$

$$x_7 \geq 0.025 \quad (26)$$

$$x_8 \geq 0.025 \quad (27)$$

The constraints owing to the coordination of relays with CTI taken as 0.6 are:

$$5.749x_3 - 5.749x_8 \geq 0.6 \quad (28)$$

$$5.584x_1 - 3.551x_4 \geq 0.6 \quad (29)$$

$$5.584x_2 - 3.551x_4 \geq 0.6 \quad (30)$$

$$4.980x_4 - 4.980x_7 \geq 0.6 \quad (31)$$

$$4.281x_1 - 2.971x_4 \geq 0.6 \quad (32)$$

$$4.980x_3 - 4.980x_8 \geq 0.6 \quad (33)$$

$$10.035x_8 - 3.297x_5 \geq 0.6 \quad (34)$$

$$10.035x_8 - 3.297x_6 \geq 0.6 \quad (35)$$

4.1.2. Application of RTO

As discussed in Section 2, to apply RTO to this problem, the objective function was first converted into an unconstrained optimization problem by integrating the relay constraint into the objective function. The constraints pertaining to the MOT of the relay have the ability to alarm the lower bounds of the RTO. The TMS values of all the relays varies from 0.025 to 1.2. The constraint owing to the MOT of the relay was taken care of defining the lower bounds of the variable in RTO. The lower and upper limit of all TMS of the relay is considered as 0.025 to 1.2. The constraints pertaining to coordination of the relay depicted by Equations (28)–(35) were incorporated in the objective function by means of a penalty scheme. In order to justify the RTO, a population size of 50 (i.e., candidate solutions, eight designed variables x_1 to x_8, and 200 iterations as meeting criterion) has been assigned to the coordination problem. The population was conveyed to the fitness function and the proceeding values of the objective function are perceived. As the objective function is of the minimization type, the lowest fitness value of the objective function is considered is the best solution among the population size. After running 200 iterations, the optimum values are shown in Table 3. Table 3 shows the optimal TMS values achieved by the proposed RTO method for this case and has been compared with the literature. Figure 6 shows the objective function values achieved during the course of the simulation for the best candidate solution in each iteration, and this figure shows that the convergence rate is faster and achieves satisfactory results in less iterations compared to other techniques cited in the literature. This figure also shows that $R_n = 0.4$, $R_r = 0.3$, and $R_c = 0.3$ are the feasible values for the RTO technique. These parameter values are used for the rest of the case studies presented in this paper as it gives an optimal solution compared with other adjustable parameters of the RTO algorithm. Figure 7 shows the graphical representation of the optimized total operating time, *Top* (z), and demonstrates that the total operating time is minimized up to the optimum value. The optimal value of objective is found to be 26.681 s by considering a CTI of 0.6 s, which is obtained in a fewer number of simulations.

Table 3. Optimal TMS for Case 1.

TMS	GA [36] ($\geq$0.6)	RTO ($\geq$0.6)
TMS 1	0.2975	0.2521
TMS 2	0.2975	0.2521
TMS 3	0.2270	0.2000
TMS 4	0.1730	0.1510
TMS 5	0.0607	0.0303
TMS 6	0.0607	0.0303
TMS 7	0.0402	0.0250
TMS 8	0.1129	0.0800
Top (z)	*31.883*	*26.681*

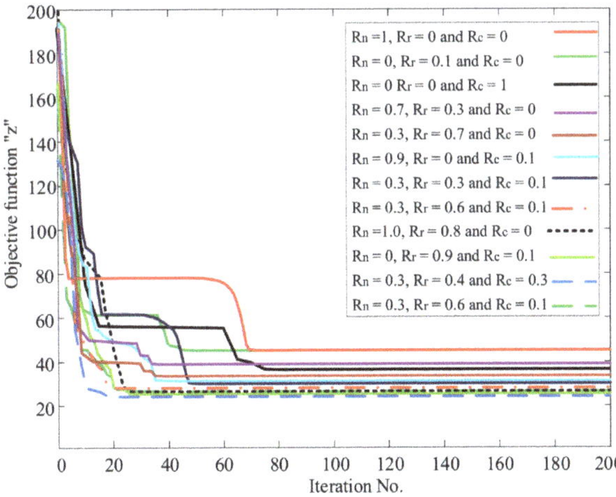

Figure 6. Convergence characteristic of Case 1 for different adjustable parameters to the RTO algorithm.

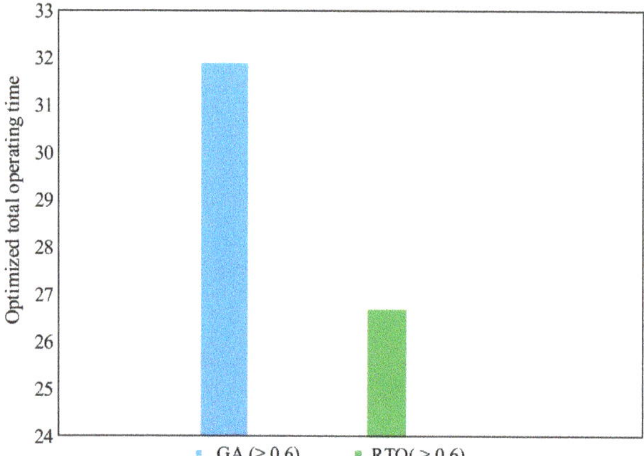

Figure 7. Optimized total operating time with the literature for Case 1.

The values depicted in Table 3 illustrate that the RTO technique gives an ideal and best-streamlined aggregate working time (i.e., total operating time) up to the ideal optimum value. The optimum values guarantee that the relay will operate in the minimum time to detect a fault at any location. All the optimal values obtained by RTO satisfy all the coordination constraints.

4.1.3. Comparison of RTO with the GA Algorithm

The results obtained by using the RTO algorithm are compared with the genetic algorithm, as given in Table 3. The RTO outperforms the genetic algorithm for obtaining the optimum TMS values and gives the advantage of 5.20198 s over the GA with the same initial conditions as supposed for this case study. The proposed RTO performs outstandingly over GA, gives an advantage in total net time gain, minimizes the total operating time up to the optimum value, and maintains proper coordination during a fault condition. In this condition, the results obtained by RTO for all the relays will satisfy the coordination constraints. Furthermore, no violation has been found regarding the coordination constraints.

4.2. Case Study 2

A multi-loop system, as appeared in Figure 8, with six overcurrent relays and with inconsequential line charging admittances are considered. Four faults are taken into consideration in the midst of the lines, i.e., A, B, C, and D.

4.2.1. Mathematical Modelling of the Problem Formulation

Table 4 demonstrates the line information of the given system. The primary/backup pair of relays for the system are shown in Table 5. For this delineation, four fault areas are taken, as shown in Figure 8. The CT proportions and plug setting are given in Table 6. For various fault areas, the current seen by relays and the a_ρ constant are given in Table 7.

Table 4. Line data for Case 2.

Line	Impedance (Ω)
1 and 2	0.08 +j1
2 and 3	0.08 + j1
1 and 3	0.16 + j2

Table 5. Primary and backup relationships of the relays for Case 2.

Fault Point	Primary Relay	Backup Relay
A	1, 2	-, 4
B	3, 4	1, 5
C	5, 6	-, 3
D	3, 5	1, -

- Indicates no backup relay.

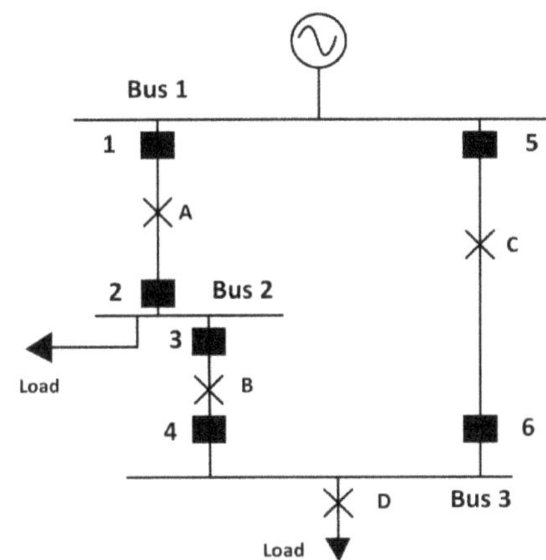

Figure 8. A single end-loop distribution system.

Table 6. CT ratios and plug settings of the relays for Case 2.

Relay	CT Ratio	Plug Setting
1	1000/1	1
2	300/1	1
3	1000/1	1
4	600/1	1
5	600/1	1
6	600/1	1

Table 7. a_ρ constants and relay currents for Case 2.

Fault Point		Relay					
		1	2	3	4	5	6
A	I_{relay}	6.579	3.13	–	1.565	1.565	–
	a_ρ	3.646	6.065	–	15.55	15.55	–
B	I_{relay}	2.193	–	2.193	2.193	2.193	–
	a_ρ	8.844	–	8.844	8.844	8.844	–
C	I_{relay}	1.096	–	1.096	–	5.482	1.827
	a_ρ	75.91	–	75.91	–	4.044	11.539
D	I_{relay}	1.644	–	1.644	–	2.741	–
	a_ρ	13.99	–	13.99	–	6.872	–

In this representation, the entire number of constraints that arise are 11. Six constraints are attributable to limits on relay working time, and five constraints are by reason of coordination criteria. The MOT of each relay is 0.1, and the CTI is considered as 0.3 s. The TMSs of all six relays are x_1–x_6.

The problem for optimization can be demonstrated as:

$$minz = 102.404x_1 + 6.0651x_2 + 98.758x_3 + 24.403x_4 + 35.319x_5 + 11.539x_6 \tag{36}$$

The constraints owing to the MOT of the relays are:

$$3.646x_1 \geq 0.1 \tag{37}$$

$$6.055x_2 \geq 0.1 \tag{38}$$

$$8.844x_3 \geq 0.1 \tag{39}$$

$$8.844x_4 \geq 0.1 \tag{40}$$

$$4.044x_5 \geq 0.1 \tag{41}$$

$$11.539x_6 \geq 0.1 \tag{42}$$

Hence, the constraints mentioned in Equations (38)–(42) violate the constraints of the minimum value of the TMS. Hence, these constraints are reconstructed as:

$$x_2 \geq 0.025 \tag{43}$$

$$x_3 \geq 0.025 \tag{44}$$

$$x_4 \geq 0.025 \tag{45}$$

$$x_5 \geq 0.025 \tag{46}$$

$$x_6 \geq 0.025 \tag{47}$$

The constraints owing to the coordination of relays with the CTI taken as 0.3 are:

$$15.55x_4 - 6.065x_2 \geq 0.3 \tag{48}$$

$$8.844x_1 - 8.844x_3 \geq 0.3 \tag{49}$$

$$8.844x_5 - 8.844x_4 \geq 0.3 \tag{50}$$

$$75.91x_3 - 11.53x_6 \geq 0.3 \tag{51}$$

$$13.998x_1 - 13.998x_3 \geq 0.3 \tag{52}$$

4.2.2. Application of RTO

The objective function was evaluated using the proposed RTO calculation with same parameters elucidated from delineation 1. The optimal values of TMS gained are given in Table 8, which shows that the RTO algorithm gives ideal and best values and that it optimized the total operating time to optimal values. The optimal values ensure that the relay will take the least possible time in the system for a fault at any area. The time taken by Relay 1 to work is the base for a fault at point A and will set aside the most noteworthy time for a relay at point C. This is alluring in light of the fact that, for a fault at point A, relay 1 is first to work while, for a fault at point C, Relay 6 should get first chance to work if it fails. Relay 3 should take over tripping action and, if Relay 3 in a similar manner fails to work, then Relay 1 ought to expect control over the tripping action. The objective function value achieved over the span of the simulation for the best applicant arrangement in every cycle is shown in Figure 9. This demonstrates that the convergence is speedier and obtained a satisfactory value in fewer iterations. The optimum value of the objective function is observed to be 11.93 by considering

CTI being taken as 0.3 s. Figure 10 demonstrates the optimized total operating time accomplished by the proposed algorithm with other methods mentioned in the literature

Table 8. Optimal TMS for Case 2.

TMS	CGA [37] ($\geq$0.3)	FA [38] ($\geq$0.3)	CFA [38] ($\geq$0.3)	CPSO [40] ($\geq$0.3)	RTO ($\geq$0.3)
TMS 1	0.0765	0.027	0.027	0.0589	0.0590
TMS 2	0.034	0.130	0.221	0.0250	0.0250
TMS 3	0.0339	0.025	0.025	0.0250	0.0250
TMS 4	0.036	0.025	0.025	0.0290	0.0290
TMS 5	0.0711	0.025	0.029	0.0630	0.0650
TMS 6	0.0294	0.489	0.363	0.0250	0.0250
Top (z)	*15.88*	*16.25*	*14.39*	*11.87*	*11.93*

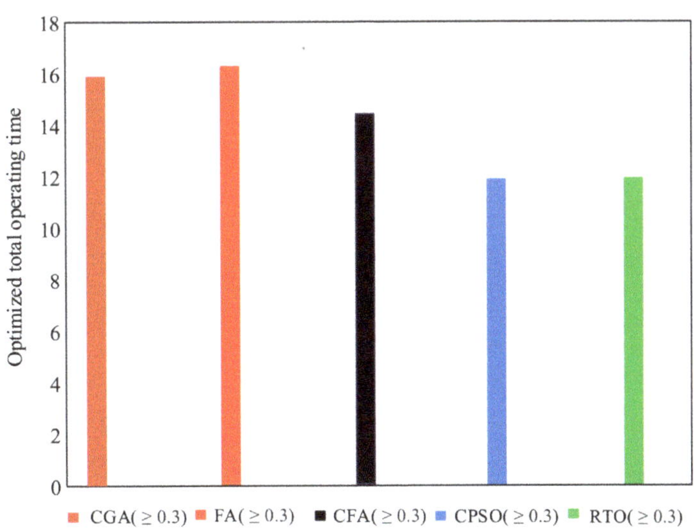

Figure 9. Convergence characteristic representation for Case 2.

Figure 10. Optimized total operating time with the literature for Case 2.

4.2.3. Comparison of RTO with Other Algorithms

To evaluate the execution of the proposed RTO algorithm, this technique was compared with the other mathematical and evolutionary techniques such as the CGA, FA, CFA, and CPSO technique accessible in the literature, as shown in Table 8. The RTO outmatched the CGA, FA, and CFA given the total net gain of times and gives advantages of 3.95, 4.32, and 2.46 s over these methods, respectively. However, in the case of CPSO, this technique almost converges to the same optimum solution as obtained by using the RTO with more computational effort. Furthermore, the RTO algorithm is better than the other techniques mentioned in the literature in terms of the nature of the results. Convergence representation and less number of iterations is required to get the ideal and best result.

4.3. Case Study 3

A parallel distribution system that is sustained from a singular end with seven overcurrent relays is shown in Figure 11. The CT ratio and PS of the relays are shown in Table 9. Four faults streams are constrained in the midst of the lines, i.e., A, B, C, and D. The primary/backup pair of relays for the given system are shown in Table 10.

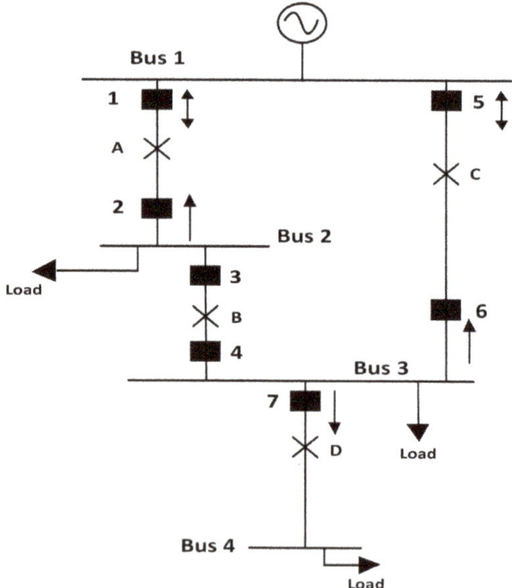

Figure 11. A single-end, multi-parallel feeder distribution system.

Table 9. CT ratios and plug settings of the relays for Case 3.

Relay	CT Ratio	Plug Setting
1	1000/1	0.8
2	1000/1	0.8
3	1000/1	0.8
4	1000/1	0.8
5	1000/1	0.8
6	1000/1	0.8
7	500/1	0.5

Table 10. Primary and backup relationships of the relays and maximum fault current for Case 3.

Fault Point	Primary Relay	Backup Relay	T. Fault Current (A)
A	1	-	6579
	4	2	939
B	3	1	2193
	4	5	1315.5
C	5	-	3289.5
	6	3	1096.5
D	7	3	1315.8 through relays 3 and 5
	-	5	2631.6 through relay 7

4.3.1. Mathematical Modelling of the Problem Formulation

For this case, the total number of constraints is 12. Seven imperatives arise as a result of the points of confinement of the relay action, and five prerequisites arise due to the coordination condition. The MOT of each relay is 0.1 s. The CTI is considered as 0.2 s. The TMS values of all the relays is x_1–x_7. For different fault regions, the current seen by relays and the a_ρ constant are given in Table 11.

Table 11. a_ρ constants and relay currents for Case 3.

Fault Point		Relay						
		1	2	3	4	5	6	7
A	I_{relay}	8.223	1.1737	-	2.347	-	-	-
	a_ρ	3.252	43.776	-	8.159	-	-	-
B	I_{relay}	2.741	-	5.482	3.288	1.644	-	-
	a_ρ	6.872	-	4.0444	5.811	14.01	-	-
C	I_{relay}	-	-	2.741	-	4.111	1.370	-
	a_ρ	-	-	6.872	-	4.881	22.165	-
D	I_{relay}	-	-	3.289	-	3.289	-	5.263
	a_ρ	-	-	5.809	-	5.809	-	4.145

The problem for optimization can be demonstrated as:

$$minz = 10.124x_1 + 43.776x_2 + 16.725x_3 + 13.97x_4 + 24.703x_5 + 22.165x_6 + 4.145x_7 \qquad (53)$$

The constraints owing to the MOT of the relays are:

$$3.252x_1 \geq 0.1 \qquad (54)$$

$$43.776x_2 \geq 0.1 \qquad (55)$$

$$4.044x_3 \geq 0.1 \qquad (56)$$

$$5.811x_4 \geq 0.1 \qquad (57)$$

$$4.881x_5 \geq 0.1 \qquad (58)$$

$$22.165x_6 \geq 0.1 \qquad (59)$$

$$4.145x_7 \geq 0.1 \qquad (60)$$

Hence, the constraints mentioned in Equations (55)–(60) violate the constraints of the minimum value of the TMS. Hence, these constraints are reconstructed as:

$$x_2 \geq 0.025 \qquad (61)$$

$$x_3 \geq 0.025 \qquad (62)$$

$$x_4 \geq 0.025 \tag{63}$$

$$x_5 \geq 0.025 \tag{64}$$

$$x_6 \geq 0.025 \tag{65}$$

$$x_7 \geq 0.025 \tag{66}$$

The constraints owing to the coordination of relays with the CTI taken as 0.2 are:

$$43.77x_2 - 8.159x_4 \geq 0.2 \tag{67}$$

$$6.872x_1 - 4.044x_3 \geq 0.2 \tag{68}$$

$$14.01x_5 - 5.811x_4 \geq 0.2 \tag{69}$$

$$22.165x_3 - 6.872x_6 \geq 0.2 \tag{70}$$

$$5.809x_3 - 4.145x_7 \geq 0.2 \tag{71}$$

4.3.2. Application of RTO

The objective function was solved by the proposed RTO with vague parameters. The ideal estimations of TMS and aggregate working time found are shown in Table 12. They similarly show the comparative delayed consequence of RTO with other metaheuristic and numerical procedures in the literature. In this depiction, no encroachment and miscoordination of the relays have been found. The time taken by relay R1 to work is the least for a fault at point C and is most extreme for a fault at point A. This is attractive on the grounds that, for a fault at point C, relay 1 is the first to work. Meanwhile, for a fault at point A, relay 7 ought to get the principal opportunity to work. On the off chance that it neglects to work, relay 4 should assume a control tripping activity. If relay 4 likewise neglects to work, relay 1 should assume control over the tripping activity at that point. The values are given in Table 12 ensure that the relays will work at any rate possible time for a fault whenever in the system. The objective value acquired over the span of the simulation for the best candidate course of action in each iteration is shown in Figure 12, which exhibits that the convergence is faster and get the best qualities in a lesser number of iterations. Additionally, the RTO outmatched the simplex method for this case and obtained an agreeable and better outcome compared to the simplex technique. The graphical image of the improved aggregate working time appears in Figure 13 and shows that the total operating time is streamlined up to the ideal and optimum value when compared to the simplex method, as mentioned in the literature.

Table 12. Optimal TMS for Case 3.

TMS	SM [39] ($\geq$0.2)	RTO ($\geq$0.2)
TMS 1	0.23829	0.0852
TMS 2	0.12	0.0250
TMS 3	0.36036	0.0953
TMS 4	0.0319	0.0250
TMS 5	0.02773	0.0250
TMS 6	0.025	0.0250
TMS 7	0.08	0.0250
Top (z)	*15.7068*	*5.1756*

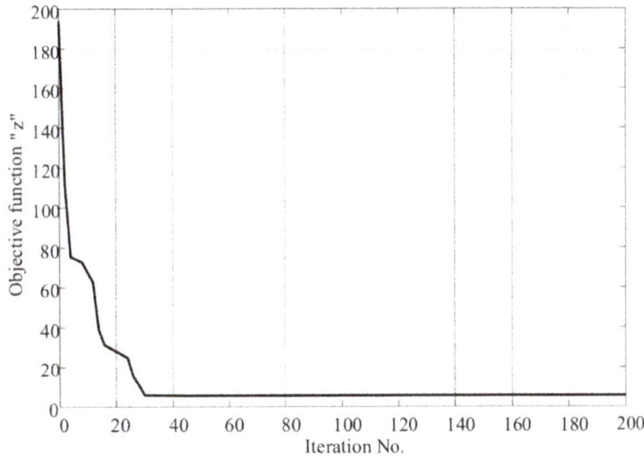

Figure 12. Convergence characteristic representation for Case 3.

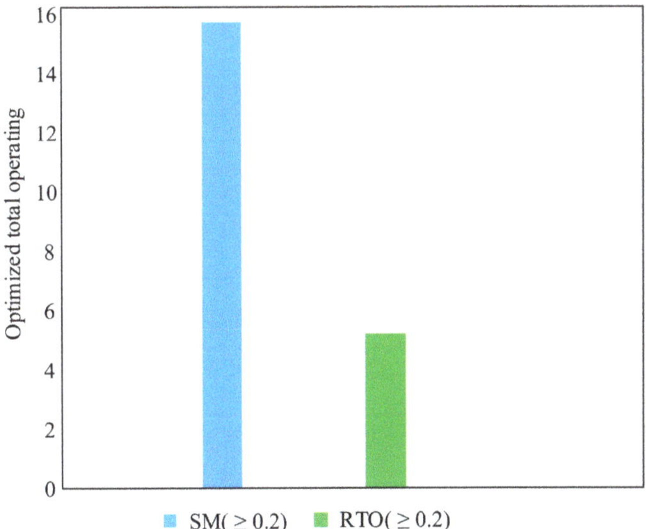

Figure 13. Optimized total operating time with the literature for Case 3.

4.3.3. Comparison of RTO with the Simplex Method

The results found using the RTO are differentiated, and the results procured by the simplex method are shown in Table 12. The RTO outmatch the simplex method in getting the optimal values of TMS and in an aggregate net gain of times. It gives a benefit of 10.5312 s over the simplex method, with the same starting parameters gathered for this contextual investigation. This gain in all out net time gain is adequate given that it is a small system. For all the ideal and optimal values for TMS, the RTO performed exceptionally compared to the simplex method, in terms of complete net time gain. Furthermore, it limits the aggregate working time up to optimal value and will keep up appropriate coordination even in a fault condition. For all the coordination conditions, the optimal value acquired by RTO for all the relays will fulfill the coordination limitations. Additionally, no infringement has been discovered with respect to the coordination requirements.

5. Conclusions

This paper suggested an RTO algorithm which simulates plant roots searching for water under the ground. The aforementioned RTO technique involved some parameters for tuning and, thus, is easy to implement. The coordination problem of the overcurrent relay utilizes the RTO algorithm for different test systems to assess the execution of the proposed RTO algorithm. The competence of the proposed RTO algorithm has been confirmed and tried through its application on various single- and multi-loop systems by analyzing its superiority and contrasting its execution with the CFA, FA, CPSO, CGA, GA, and the simplex method published algorithms. The simulation results of the RTO algorithm efficiently minimize all three models of the problem. The performance of the RTO can be seen from the optimized minimum objective function values and TMS, of each relay in the systems achieved by RTO for each case studies. In case 1 the objective function value is minimized up to optimum value by RTO and gives an advantage in total net gain in time of 5.20198 s over GA. while in case 2 and 3 the RTO gives a total net gain in time 3.95 s, 4.32 s, 2.46 s, and 10.5312 s over CGA, FA, CFA, and simplex method. The RTO algorithm gives a new seeking approach for elucidation, as one of its qualities is the extensive field of research attributable to the activities of the roots. The simulation results uncover the predominance of the proposed RTO algorithm in taking care of the overcurrent relay coordination problem. In future work, RTO will be applied to solve the problem of overcurrent relay coordination of higher and complex buses in the electrical power system.

Author Contributions: A.W. conceived and designed the algorithm. A.W., S.G.F., and C.-H.K. designed and performed the experiments. A.W., T.K., and J.Y. wrote the paper. A.W., S.G.F., and S.-B.R. formulated the mathematical model. S.-B.R. and Z.W.G. supervised and finalized the manuscript for submission.

Acknowledgments: This research was supported by Korea Electric Power Corporation, grant number (R17XA05-38).

Conflicts of Interest: The authors declare no conflict of interest.

References

1. Urdaneta, A.J.; Nadira, R.; Jimenez, L.G.P. Optimal coordination of directional overcurrent relays in interconnected power systems. *IEEE Trans. Power Deliv.* **1988**, *3*, 903–911. [CrossRef]
2. Birla, D.; Maheshwari, R.P.; Gupta, H.O.; Deep, K.; Thakur, M. Application of random search technique in directional overcurrent relay coordination. *Int. J. Emerg. Electr. Power Syst.* **2006**, *7*. [CrossRef]
3. Ehrenberger, J.; Švec, J. Directional Overcurrent Relays Coordination Problems in Distributed Generation Systems. *Energies* **2017**, *10*, 1452. [CrossRef]
4. Ates, Y.; Boynuegri, A.; Uzunoglu, M.; Nadar, A.; Yumurtacı, R.; Erdinc, O.; Paterakis, N.; Catalão, J. Adaptive protection scheme for a distribution system considering grid-connected and islanded modes of operation. *Energies* **2016**, *9*, 378. [CrossRef]
5. Núñez-Mata, O.; Palma-Behnke, R.; Valencia, F.; Jiménez-Estévez, P.M.G. Adaptive Protection System for Microgrids Based on a Robust Optimization Strategy. *Energies* **2018**, *11*, 308. [CrossRef]
6. Elrafie, H.B.; Irving, M.R. Linear programming for directional overcurrent relay coordination in interconnected power systems with constraint relaxation. *Electr. Power Syst. Res.* **1993**, *27*, 209–216. [CrossRef]
7. Aghdam, T.S.; Hossein, K.K.; Hatem, Z. Optimal Coordination of Double-Inverse Overcurrent Relays for Stable Operation of DGs. *IEEE Trans. Ind. Inform.* **2018**. [CrossRef]
8. Aghdam, T.S.; Karegar, H.K.; Zeineldin, H.H. Transient Stability Constrained Protection Coordination for Distribution Systems with DG. *IEEE Trans. Smart Grid* **2017**. [CrossRef]
9. Ojaghi, M.; Mohammadi, V. Use of Clustering to Reduce the Number of Different Setting Groups for Adaptive Coordination of Overcurrent Relays. *IEEE Trans. Power Deliv.* **2018**, *33*, 1204–1212. [CrossRef]
10. Braga, A.S.; Saraiva, J.T. Coordination of overcurrent directional relays in meshed networks using the Simplex method. In Proceedings of the 8th Mediterranean Electrotechnical Conference on Industrial Applications in Power Systems, Computer Science and Telecommunications (MELECON'96), Bari, Italy, 16 May 1996; Volume 3.

11. Abyaneh, H.A.; Keyhani, R. Optimal co-ordination of overcurrent relays in power system by dual simplex method. In Proceedings of the Australasian Universities Power Engineering Conference (Aupec'95), Nedlands, Australia, 27–29 September 1995; Volume 3.

12. Abdelaziz, A.Y.; Talaat, H.E.A.; Nosseir, A.I.; Hajjar, A.A. An adaptive protection scheme for optimal coordination of overcurrent relays. *Electr. Power Syst. Res.* **2002**, *61*, 1–9. [CrossRef]

13. Laway, N.A.; Gupta, H.O. A method for adaptive coordination of overcurrent relays in an interconnected power system. In Proceedings of the 1993 Fifth International Conference on Developments in Power System Protection, York, UK, 30 March–2 April 1993.

14. Sulaiman, M.; Waseem; Muhammad, S.; Khan, A. Improved Solutions for the Optimal Coordination of DOCRs Using Firefly Algorithm. *Complexity* **2018**, *2018*, 7039790. [CrossRef]

15. Mansour, M.M.; Mekhamer, S.F.; El-Kharbawe, N. A modified particle swarm optimizer for the coordination of directional overcurrent relays. *IEEE Trans. Power Deliv.* **2007**, *22*, 1400–1410. [CrossRef]

16. Zeineldin, H.H.; El-Saadany, E.F.; Salama, M.M.A. Optimal coordination of overcurrent relays using a modified particle swarm optimization. *Electr. Power Syst. Res.* **2006**, *76*, 988–995. [CrossRef]

17. Razavi, F.; Abyaneh, H.A.; Al-Dabbagh, M.; Mohammadi, R.; Torkaman, H. A new comprehensive genetic algorithm method for optimal overcurrent relays coordination. *Electr. Power Syst. Res.* **2008**, *78*, 713–720. [CrossRef]

18. Kim, C.H.; Khurshaid, T.; Wadood, A.; Farkoush, S.G.; Rhee, S.B. Gray Wolf Optimizer for the Optimal Coordination of Directional Overcurrent Relay. *J. Electr. Eng. Technol.* **2018**, *13*, 1043–1051.

19. Thangaraj, R.; Pant, M.; Deep, K. Optimal coordination of over-current relays using modified differential evolution algorithms. *Eng. Appl. Artif. Intell.* **2010**, *23*, 820–829. [CrossRef]

20. Bouchekara, H.R.E.H.; Zellagui, M.; Abido, M.A. Optimal coordination of directional overcurrent relays using a modified electromagnetic field optimization algorithm. *Appl. Soft Comput.* **2017**, *54*, 267–283. [CrossRef]

21. Alipour, M.; Teimourzadeh, S.; Seyedi, H. Improved group search optimization algorithm for coordination of directional overcurrent relays. *Swarm Evolut. Comput.* **2015**, *23*, 40–49. [CrossRef]

22. Brown, K.; Tyle, N. An expert system for over current protective device coordination. McGraw Edison power system division. In Proceedings of the IEEE Rural Electric Power System Conference, Charleston, SC, USA, April 1986; pp. 20–22.

23. Lee, S.J.; Yoon, S.H.; Yoon, M.C.; Jang, J.K. An expert system for protective relay setting of transmission systems. *IEEE Trans. Power Deliv.* **1990**, *5*, 1202–1208.

24. Decher, G.; Hong, J.-D. Buildup of ultrathin multilayer films by a self-assembly process, 1 consecutive adsorption of anionic and cationic bipolar amphiphiles on charged surfaces. *Macromol. Symp.* **1991**, *46*, 321–327. [CrossRef]

25. Wang, J.; Trecat, J. RSVIES—A relay setting value identification expert system. *Electr. Power Syst. Res.* **1996**, *37*, 153–158. [CrossRef]

26. Abyane, H.A.; Faez, K.; Karegar, H.K. A new method for overcurrent relay (O/C) using neural network and fuzzy logic. In Proceedings of the IEEE Region 10 Annual Conference, Speech and Image Technologies for Computing and Telecommunications (TENCON'97), Brisbane, Australia, 4 December 1997; Volume 1.

27. Yu, J.; Kim, C.H.; Wadood, A.; Khurshiad, T.; Rhee, S.B. A Novel Multi-Population Based Chaotic JAYA Algorithm with Application in Solving Economic Load Dispatch Problems. *Energies* **2018**, *11*, 1946. [CrossRef]

28. Farkoush, S.G.; Khurshiad, T.; Wadood, A.; Kim, C.-H.; Kharal, K.H.; Kim, K.-H.; Cho, N.; Rhee, S.-B. Investigation and Optimization of Grounding Grid Based on Lightning Response by Using ATP-EMTP and Genetic Algorithm. *Complexity* **2018**, *2018*, 8261413. [CrossRef]

29. Park, J.H.; Yu, J.S.; Geem, Z.W. Genetic Algorithm-based Optimal Investment Scheduling for Public Rental Housing Projects in South Korea. *Int. J. Fuzzy Log. Intell. Syst.* **2018**, *18*, 135–145. [CrossRef]

30. Moon, Y.Y.; Geem, Z.W.; Han, G.-T. Vanishing point detection for self-driving car using harmony search algorithm. *Swarm Evolut. Comput.* **2018**, *41*, 111–119. [CrossRef]

31. Geem, Z.W.; Yoon, Y. Harmony search optimization of renewable energy charging with energy storage system. *Int. J. Electr. Power Energy Syst.* **2017**, *86*, 120–126. [CrossRef]

32. Wadood, A.; Kim, C.; Loon, T.K.; Hassan, K.; Farkoush, S.G.; Rhee, S.-B. Optimal Coordination of Directional Overcurrent Relays using New Rooted Tree Optimization Algorithm. In Proceedings of the International Conference on Information, System and Convergence Applications (ICISCA/ICW 2018), Bangkok, Thailand, 31 January–2 February 2018.

33. Chattopadhyay, B.; Sachdev, M.S.; Sidhu, T.S. An on-line relay coordination algorithm for adaptive protection using linear programming technique. *IEEE Trans. Power Deliv.* **1996**, *11*, 165–173. [CrossRef]

34. Wadood, A.; Kim, C.-H.; Farkoush, S.G.; Rhee, S.B. An Adaptive Protective Coordination Scheme for Distribution System Using Digital Overcurrent Relays. *Korean Inst. Illum. Electr. Install. Eng.* **2017**, *5*, 53.

35. Labbi, Y.; Attous, D.B.; Gabbar, H.A.; Mahdad, B.; Zidan, A. A new rooted tree optimization algorithm for economic dispatch with valve-point effect. *Int. J. Electr. Power Energy Syst.* **2016**, *79*, 298–311. [CrossRef]

36. Bedekar, P.P.; Bhide, S.R.; Kale, V.S. Optimum coordination of overcurrent relays in distribution system using genetic algorithm. In Proceedings of the International Conference on Power Systems (ICPS'09), Kharagpur, India, 27–29 December 2009.

37. Bedekar, P.P.; Bhide, S.R. Optimum coordination of overcurrent relay timing using continuous genetic algorithm. *Expert Syst. Appl.* **2011**, *38*, 11286–11292. [CrossRef]

38. Gokhale, S.S.; Kale, V.S. An application of a tent map initiated Chaotic Firefly algorithm for optimal overcurrent relay coordination. *Int. J. Electr. Power Energy Syst.* **2016**, *78*, 336–342. [CrossRef]

39. Bedekar, P.P.; Bhide, S.R.; Kale, V.S. Optimum coordination of overcurrent relay timing using simplex method. *Electr. Power Compon. Syst.* **2010**, *38*, 1175–1193. [CrossRef]

40. Wadood, A.; Kim, C.H.; Khurshiad, T.; Farkoush, S.G.; Rhee, S.B. Application of a Continuous Particle Swarm Optimization (CPSO) for the Optimal Coordination of Overcurrent Relays Considering a Penalty Method. *Energies* **2018**, *11*, 869. [CrossRef]

Article

Designing a Waterless Toilet Prototype for Reusable Energy Using a User-Centered Approach and Interviews

Hyun-Kyung Lee

Division of General Studies, UNIST, Ulsan 44919, Korea; hlee@unist.ac.kr

Received: 24 January 2019; Accepted: 25 February 2019; Published: 4 March 2019

Abstract: User-oriented community engagement can reveal insights into ways of improving a community and solving complex public issues, such as natural resource scarcity. This study describes the early process of co-designing a novel, waterless toilet to respond to the water scarcity problem in the Republic of Korea. It presents how we designed a toilet focusing on three factors—a sanitization function, an ergonomic posture, and clean aesthetics—by conducting focus group interviews as part of a user engagement approach to understand what community users want from a toilet and ways of improving their toilet experiences. The results not only supported the development of an experiential service design project to raise community awareness of water scarcity but also supported scientists and engineers in experimenting with and developing new technologies by collaborating closely with designers.

Keywords: design prototype; toilet design; collaboration with scientists; interdisciplinary convergence; natural resource; feces

---

1. Introduction

Community engagement is a suitable approach for solving complex public issues, such as natural resource scarcity. It may also reveal what people already know about an issue and provide insights that could support the creation of a better natural resource community. This study describes the fuzzy front end (FFE) stage of designing a novel, waterless toilet prototype, the first collaborative research project between designers and scientists to respond to the water scarcity problem in South Korea. The aim of this collaboration was not only to raise community awareness of water scarcity, but also to save water. The co-design of the waterless toilet at the FFE followed a community-engaged research approach that influenced the technology push and supported "silent design": that is, design decisions made by non-designers [1]. This collaboration project sought to understand what users want in a toilet and to improve the toilet experience. Unpredictably, the approach led to the creation of an experiential service design project to raise community awareness of water scarcity.

2. Background: Water Scarcity

Water is one of the most vital resources for human viability. However, there is not enough for the needs of everyone in the globe. Due to population growth, many countries face water shortages; thus, they need to prepare for future problems [2]. Water scarcity also results in low economic growth. In 2012, the Organisation for Economic Co-operation and Development (OECD) recommended that both OECD and non-OECD countries reduce water risks, such as water shortages (including droughts), inadequate water quality, non-accessible water, and anything that undermines the resilience of freshwater. Yet, all countries face different water scarcity problems.

According to the United Nations Environment Programme [3], the rapid economic growth and high population density of the Republic of Korea have caused water shortages and freshwater scarcity.

This has long been—and still remains—a critical challenge. Reports indicate that current aqua-ecosystem protection mechanisms are insufficient [3]. The Ministry of Land, Transport and Maritime Affairs (2011) reported that only 26% of water in Korea is fit for use and that each Korean person uses only one-sixth of the global average for individual water consumption. Hence, the nation is considered a water-stressed nation.

Water in the Republic of Korea is contaminated for three reasons: (1) damage to the nation's hydro-ecology caused by construction near rivers, which has decreased limnobios, animals, and freshwater plants; (2) soil runoff; and (3) the eutrophication of rivers due to an increasing inflow of phosphorus into rivers [4]. According to the previous literature, people consider treated wastewater from sewage treatment plants to be clean; however, attached algae are widely visible, largely due to phosphorus discharged from farmlands during the summer flood season. More critically, the shallow depth and slow current speed of the river increase detention time, increasing nutrient salts during annual drought seasons. Since phosphorus is not filtered from sewage, it grows too much. It originally comes from toilets, and some toilets are installed in areas that lack sewage. Thus, it is important to remove phosphorus from toilets. This can be accomplished by a sewage treatment plant.

In 2015, the Ministry of Science, ICT (Information and Communications Technologies), Future Planning (MSIP), initiated the Convergence Research Centre (CRC) project, a cross-disciplinary research project that connects the arts, design, and science to prepare for future sustainable energy. Solving the water scarcity problem is one of several seven-year research projects. The collaboration team comprised two design researchers, one philosopher, seven different disciplinary scientists (e.g., water, climate change, and urban environment), and industry partners (e.g., different groups of natural artists, digital artists, graphic designers, and engineers). The team is planning to build a temporary research lab to research these issues at the UNIST (Ulsan National Institute of Science and Technology) in the Republic of Korea. The research lab building will be two stories high and approximately 99 m^2.

3. Waterless Toilet

A family of four uses an average of 225 L of water per day. This represents 27% of household water consumption [5]. Domestic houses use 65.9% of total water [6] and represent one of the main causes of contaminated water in Korea. To remove phosphorus, water scientists from the Department of Urban and Environmental Science initiated a waterless toilet project.

The conventional sanitation system, the flush toilet, has improved public health and protected humans from waterborne diseases. The "drop-and-discharge" approach is a convenient and comfortable solution for disposing human excreta because it rapidly removes hazardous and unpleasant matter from households [7]. However, the complete disconnect between people and their excreta causes environmental problems. People discharge their excreta unconsciously, meaning that our environment has been forced to receive the massive flow of contaminants. In particular, algal blooming and red-tide problems occur in many countries, even developed countries with well-established wastewater treatment systems. Waterless toilets that dispose of excreta in a sanitary way could reduce the high level of clean water usage. Returning sanitized dried feces to the land may also improve the soil environment. Thus, a waterless toilet team—one of the CRC project teams—has begun to investigate alternative toilet systems. The team has explored ways to develop a prototype to examine the effectiveness of eliminating phosphorus from the fuzzy front end (FFE) stage, which is considered the earliest stage of the new product development process [8].

A prototype can be a tool for solution generation or evaluation, as well as a vehicle for team collaboration [9,10]. Supporting explorations that allow users to innovate new patterns of use in the field and over lengthy periods may provide designers insights into the domestication of radically new concepts [11].

4. User-Centered Research Approach

4.1. The FFE Process

This project is characterized by a radical technology push, such that engineers in the field develop a set of research routines based on their beliefs about what is feasible or worth trying [12–14]. However, according to NESTA [15], so-called "front-end" research into technological development, market trends, and consumer needs are now staples in design industry operations. Numerous studies identify communication as critically important because this phase allows for modifications, reorientations, and radical changes in new product planning [16]. FFE is the least expensive stage of a project [16]. This collaboration team, therefore, communicated once every two or three weeks regarding how a prototype, by developing technology to remove phosphorous from a waterless toilet, should be designed. In order to clarify the product's concept during the FFE stage, which is known to be complicated (most companies fail to have clear product definitions [17]), we reviewed the literature on new product development (NPD) during the FFE stage. Based on a theory of marketing, operation management, and engineering design, Krishnan and Ulrich define product development as "the transformation of a market opportunity and a set of assumptions about product technology into a product available for sale" [18]. Similarly, Ulrich and Eppinger define product development as "the set of activities beginning with the perception of a market opportunity and ending in the production, sales, and delivery of a product" [19]. A successful product may be viewed as a solution to customers' demands [20]. In order to design elegant and efficient solutions to customers' needs, two different types of information must be combined: "need information" (what users need) and "solution information" (how products are built) [20]. The National Endowment for Science and Technology and the Arts (NESTA) has proposed that researching people's needs, tastes, and preferences is critical for shaping new products and services [15]. However, though consumer needs should be the focal point of any design process, many organizations neglect them [21]. As a result, many design processes fail to reach their target consumers or end users [22,23]. In addition, due to the rapidity of modern technological development, many users find it difficult to directly specify their needs regarding a particular product, service, or experience [24,25]. For this reason, Borja de Mozota recommends employing user-centered and informed design research methods to support the development of products and services from the beginning so not to fail to find out what users need [26].

To design something that is meaningful for users, it is important to understand users' experiences, perceptions, and ways of making sense of things [27]. Having a thorough understanding of one's target users and their needs is necessary to be successful. Mossberg found that the growth of industry through customer experience is reflected in new patterns of consumption, new demands, and new technology [28]. He insisted that a customer's experience is a blend of many elements that have come together to involve consumers emotionally, physically, intellectually, and spiritually in an ethnographic approach. Similarly, Pine and Gilmore propose that combining NPD with customer experience creates more profit than applying NPD on its own because, in the combined approach, every touch point with the customer serves as a market opportunity [29]. Several successful organizations that regard customer experience as an aspect of economic value have been evaluated [30]. These companies position experience as a planned journey with multiple touch points, such that the packaged experience defines the characteristics of a product, service, or brand [31]. Taking the above into consideration, we aimed to identify and learn about community members who would be potential future users of our waterless toilet. Specifically, we followed a reflective problem-solving approach [32] to design prototyping. Figure 1 presents how user-centered research influenced the design of our waterless toilet.

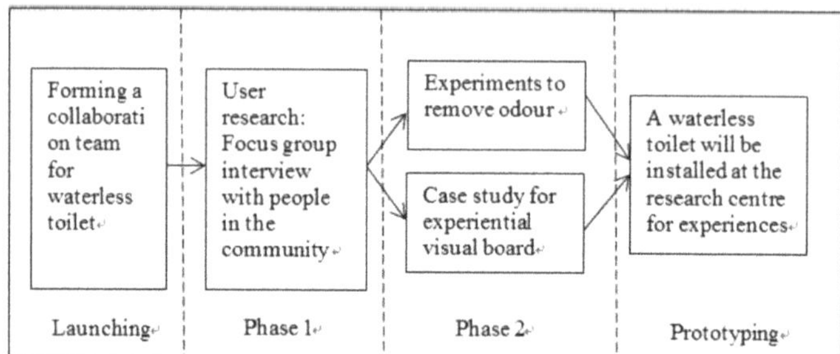

Figure 1. Co-designing a waterless toilet using the fuzzy front end (FFE) process.

4.1.1. Phase 1: Focus Group Interviews

We conducted focus group interviews to investigate user perceptions and understandings of the water scarcity problem and a waterless prototype.

Sampling Criteria

The study aimed to understand the people in the UNIST campus community where the test-bed research lab will be built. A total of 54 participants (15 females and 39 males)—all of whom had used public toilets and toilets in their homes and who would be potential users of the waterless toilet prototype—were involved in the focus group interviews. There were nine groups, each of which comprised six UNIST undergraduate students majoring in different areas of engineering (e.g., life science, chemical engineering, physics, computer engineering, electronic engineering, mechanical engineering, and engineering management).

Process

To ensure that all of the key issues were covered, semi-structured interview questions were used. The participants were asked about (1) their awareness of water scarcity in Korea, (2) their previous experiences attempting to save water, (3) how they could be motivated to save water, (4) their perceptions of a waterless toilet, and (5) their dream toilet. They were also asked to share their thoughts and ideas about how to save water. All interviews were audio-recorded and recorded via memo form that I used in collecting and writing about my observations, which were later analyzed using thematic analysis.

Results

Though the participants in each group had different majors, all shared common ideas and perceptions about each question. The thematic analysis revealed three themes: a disbelief about water scarcity, a motivation to save water, and an experience requirement for the waterless toilet.

(1) Disbelief about water scarcity: Everyone had heard that the Republic of Korea lacked water since they were children. However, most did not think that this was true, since the water quality in Korea is good and since they could access sufficient water to meet their needs anytime and anywhere. One individual thought that the push to save water (i.e., the many water-saving media campaigns) was propaganda. Only a few people had actually tried to save water through such methods as "using a cup to brush their teeth," "filling a water basin to wash their faces," or, in the case of one individual, "putting a plastic bottle in the toilet water tank to flush less water."

(2) Motivation for saving water: Most participants said that they had not attempted to try to save water because they did not believe that the Republic of Korea was truly suffering from water scarcity. The participants had no direct experiences of water scarcity, and most mentioned that, since their water bills were not expensive, they were not aware of how much water they

used each day. They were from a demographic that was less likely to experience shortage or poor quality of water. As possible motivators to save water, the participants recommended water usage indicators or visual indicators of water consumption. They also suggested that more media exposure would be helpful in raising awareness of water scarcity. Lastly, they commented that financial losses would motivate them to save water. Some participants wanted tax deductions for saving water.

(3)   Experience requirement for the waterless toilet: This theme emerged from the two questions about the participants' perceptions about a waterless toilet and their dream toilets. None of the participants had previously thought about their dream toilet. However, they said that they wanted a new toilet based on their public and private toilet experiences: visual cleanness, sanitariness (including automatic cleaning around the inside of the bowl), automatic flushing after usage, no bad odor, soundproofing within a public toilet, and a comfortably warm seat cover. Since the participants had not previously considered the water scarcity problem, they explained ways in which they would improve the current style of toilet. They also said that they would try a waterless toilet if they came across it in a public restroom. Most asked for simple visual guidance on how to use a waterless toilet, which is designed to suck up feces like a "vacuum cleaner" and send it directly to the energy production system. It requires about half liters of water, which is significantly less than what a regular toilet consumes.

The results of the focus group interviews revealed that, although people had heard of the water scarcity problem, they had never before tried to save water because they could always access water easily and at their convenience. A design activity exposes new issues and information needs as the work progresses [33,34]. The aim of designing the waterless toilet prototype was to improve the situation water scarcity; however, if the community is not aware that there is a genuine water scarcity problem, they are less likely to accept a waterless toilet. Therefore, since the participants required a realistic experience to motivate them to save water, we chose to design not only a waterless toilet but also an experiential visual guideline that indicates the water scarcity problem. Moreover, the engineering team decided to remove all negative odors from the toilet. This decision was based on the user-centered approach and was considered a sociable design, or a design "for the benefit of the people who use it, taking into account their true needs and wants" (Norman 2010, 130). Thus, the co-design of the waterless toilet, which was led by a technology push, was divided into two projects to meet the needs of the people in the community who would become the future users of the toilet and participants in solving the water scarcity problem.

4.1.2. Phase 2: Case Studies for Service Design Guideline

The aim of case studies is to study good practices of visual and experiential designs capable of leading and increasing civic engagement and awareness of the need to save energy.

Criteria

In order to identify case studies capable of meeting the experiential aims of civic engagement by providing realistic visual experiences, we contacted the Korea Institute of Design Promotion (KIDP), a non-profit organization for design support under the Ministry of Trade, Industry and Energy. Though there is no service design related to water scarcity, the design policy experts and staff of the KIDP recommended three service design cases for the public sector in Korea, all involving the design and management of public campaigns to reduce energy bills and raise awareness of the need to save energy. These studies explored (1) which visual elements were strategically designed to lead or motivate civic engagement to reduce energy usage and (2) which approaches and principles were applied to the design experience to increase people's engagement.

(1)   Redesigning an apartment energy bill (http://www.slideshare.net/sdnight/ss-30524771). This case concerned one of the first service designs for a utility service in Korea. It was conducted

on an apartment town comprising 600 apartments in Bangbae Dong, Seoul, and its results were highly effective, reducing the total energy bills by 10%. Many redesigning projects graphically change a certain part of an energy bill; however, in this case, the designers and design researchers reduced the energy bill based on in-depth interviews with the community. This case study prompted other apartment towns to accept similar guidelines.

(2)   National Health Insurance Service (http://www.slideshare.net/usableweb/ss-16567992?related=1). This case involved redesigning a health check-up chart for the National Health Insurance Service at Ilsan, Myungji Hospital. The chart had been criticized for being difficult to read by people who were not medical doctors from specific areas. In other words, doctors in one medical center often struggled to understand examinations conducted in other medical centers. The redesigned check-up chart received a 95% satisfaction rating.

(3)   Changwon National Industrial complex. As foreign labor for industrial complexes has increased, many industrial safety accidents have occurred. To reduce safety accidents, which can also waste energy resources (e.g., in cases of hydrofluoric acid leaks), a user-centered approach was employed. The service design involved changing signage and installing pictograms for international laborers. This case was published online (http://economy.hankooki.com/lpage/industry/201504/e20150406173204120170.htm).

Process

Secondary data of visual campaign were collected through online case design websites (www.designdb.com) and suggested website addresses. Thematic analysis was used to identify each case's key characteristics in relation to creating civic engagement experiences, reducing energy usage, and motivating people to do things they had not done before.

Case Study Results

Though all three cases were different, they shared similar characteristics in that all employed a user-centered approach to encourage people to reduce their energy usage or do things in which they had not previously been involved. Two themes emerged.

(1)   Easy readability. Most people were interested in their health conditions or how much they were charged on their utility bill. However, previous bills had been designed with small font sizes and logos or advertisements. Based on user interviews, both bills were redesigned to use larger fonts and exclude or deemphasize less important information.

(2)   Vivid color for nudging. Both the apartment bill case and the industrial complex case used a vivid red color for warnings. If an individual used more energy than that used by the average household nationwide, he or she received a red-colored bill that was visible to every neighbor. If an individual used less energy than the national average, he or she received a green-colored bill. Average utility usage generated a yellow-colored bill. Being able to compare their consumption with that of their neighbors motivated many users to reduce their consumption. Similarly, in the case of the national industrial complex, toxic pipelines were colored red to alert international laborers who did not know Koreans to be cautious. This step reduced workplace accidents. In short, using a vivid color and improving readability increased competitiveness among residents and safety among laborers.

4.2. Influence of User Research on the Waterless Toilet Design

These case studies made it clear that a user research approach reveals opportunities to improve civic engagement. Mayhew and Bias indicated that such an approach supports higher returns in terms of increased product value [35]. The case study participants actively participated to reduce utility bills and safety accidents. They seemed to be tuned into the effort to overcome problems caused by a lack of staff [36]. Moreover, Boyle and Harris indicated that collaborating with users could guarantee that services meet their requirements. Thus, user research supports better outcomes and encourages active user engagement in self-help and positive behavioral change, thus avoiding possible challenges in the future.

In order to engage with people in the community (a research test bed at UNIST), we used focus group interviews to determine community members' needs for realistic experiences related to water scarcity and for vivid visual representations to notify/warn about utility usage, health insurance check-ups, and safety cautions. To raise awareness about water scarcity in the community, the collaboration teams recommended encouraging people to visit the research lab to examine the waterless toilet prototype. Ramaswamy and Gouillart indicated that the participatory approach could improve stakeholder engagement, and could lead to higher productivity, higher creativity, and lower costs and risks [37]. Thus, the researchers created a visual board as a service design, which served as a channel for experience prototyping [38]. After the prototype at the research lab was developed and designed, this visual board was used to support community members in discussing their experiences and perceptions on water scarcity. As the community members visited the lab, it was possible to monitor their different behaviors in a real context using several variations of observation-based ethnographic field methods [39]. It was also possible to involve the users in designing and delivering services to achieve the full benefits of co-creation [39].

5. Prototyping: Waterless Toilet Design

In designing the toilet, we focused on three factors: the sanitization function, an ergonomic posture, and cozy aesthetics.

Sanitization Function

First, from the focus group interview, we drew the conclusion that people want a clean toilet experience. Thus, we designed the prototype for sanitization (sterilization) and cleanliness. We installed ultraviolet (UV) lights on the cover so that the toilet looks sanitizable. These UV lights also disinfect and sterilize the surface of the seat and the bowl of the toilet (Figure 2). Additionally, since the word "toilette" means "dressing room" in French, the seat cover was designed to resemble a dressing room vanity chair, which was intended to make people feel warm. The toilet hip area was designed with enough space for anyone to sit comfortably. Ultimately, the toilet was designed to provide not only a clean experience, but also a restful one.

Figure 2. Prototype of the waterless toilet.

Ergonomic Posture

Second, the toilet was designed in an ergonomic manner to support proper posture for defecating. The seat can be adjusted (raised or lowered) to an appropriate angle to fully relax the muscle around the colon and support quick and easy defecation. Once a user sits down on the toilet, the back side of seat tilts slightly, raising the user's knees. This encourages users to maintain an optimal posture for relaxing the muscle around the colon. The concept for this posture came from the Paimio Chair, designed by Alvar Aalto (Figure 3) for patients in a tuberculosis sanatorium in Finland. When the patient sits down and leans on the chair, the patient's posture changes the angle of the chair, helping the patient breathe better and comfortably arranging his or her body organs.

Figure 3. The Paimio Chair designed by Alvar Aalto in Finland.

The waterless toilet is also designed ergonomically with respect to the angle of the sitting posture. Although the tilted seat relaxes the muscle around the colon, it also makes it difficult for users to stand up (compared to the upright posture of a traditional toilet). Though this may not pose a significant problem for young and healthy people, it could encumber older, disabled, or injured people. Therefore, we installed a spring under the back side of the seat that users can activate by pushing back slightly when they need to stand up. Figures 4 and 5 present suboptimal and optimal posture angles defecation.

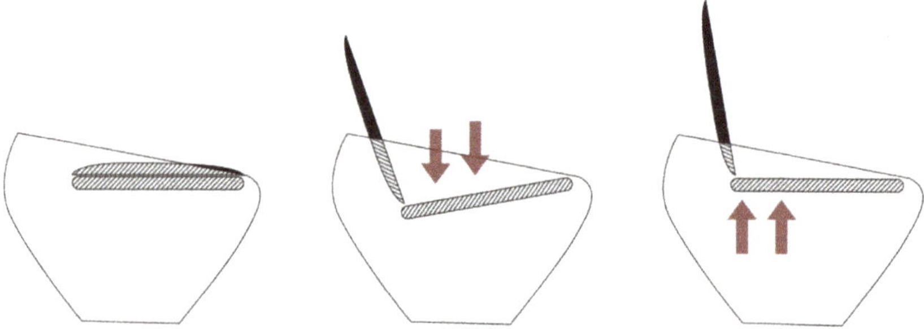

Figure 4. Working process of the seat of the waterless toilet prototype.

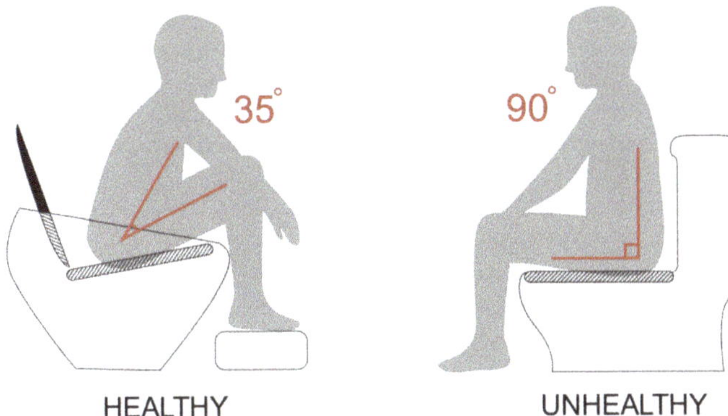

Figure 5. The proper posture and angle to relax a human's intestines.

Cozy Aesthetics

Thirdly, the appearance of the toilet prototype was inspired by the white porcelain of the Yi Dynasty (Korean imperial household, called the Joseon household) (Figure 6).

Figure 6. Korean Yi Dynasty porcelain.

Toilet shapes have not changed very much since their first development in the 16th century. However, this new type of toilet offers several improvements. First, unlike a traditional toilet, which has a tapered bowl shape to support a suction mechanism located at its bottom back side, our proposed toilet does not require centrifugal force or a vacuum system to suction the feces. Therefore, the waterless toilet can be slimmer or wider. We chose a shape that was cozy and familiar. Second, the white porcelain of the Yi Dynasty suggests familiar characteristics of interior objects that already play a role in our everyday lives. Figure 6 presents the aesthetics of the waterless toilet. Users think this is a great idea that has Korean tradition and culture embedded in it.

6. Conclusion

This article describes the FFE process of designing a waterless toilet to combat the problem of water scarcity in the Republic of Korea. The waterless toilet prototype was co-designed based on user perceptions, a technology push, and the use of a radical technology to remove phosphorus from toilets. Through focus group interviews, we identified several key factors regarding water scarcity and users' perceptions of a waterless toilet: (1) disbelief concerning the state of water scarcity in Korea, (2) a motivation to save water, and (3) experience requirements concerning the waterless toilet.

We observed that people in Korea generally do not seriously consider Korea's water scarcity problem. However, if they experience water scarcity visually and realistically, they may be more likely to try to save water. We also found that people want a toilet to be sanitary, odorless, clean, and comfortable. These findings helped to reveal what users want and need to improve their toilet experience, and they guided the direction of the prototype development to be more user-oriented and less rushed (i.e., due to the technology push).

Based on our user research, we designed our waterless toilet focusing on three main factors: a sanitization function, an ergonomic posture, and cozy aesthetics. First, based on the focus group interview, we concluded that people want a clean toilet experience. The prototype was therefore designed with UV lights on its cover for sanitization (sterilization) and cleanliness. Second, the toilet was designed in an ergonomic manner to allow people to sit in a posture appropriate for defecating. Specifically, the seat can be lowered and adjusted to the optimal angle to relax the muscle around the colon. Thirdly, the appearance of the toilet prototype was inspired by the white porcelain of the Yi Dynasty and includes familiar characteristics of interior elements and everyday objects.

The researchers' user-centered and community-focused approach helped this collaborative research project gain new information for further improvements. The approach used in this study can be employed as a guideline for researching and solving other public problems. Engaging people in the community supports the identification of both current and future public problems. Furthermore, responding to potential users' opinions, recommendations, and insights encourages positive participation and thus a better public environment. Further studies evaluating the waterless toilet prototype with the aid of community members are needed to improve the toilet experience and to increase awareness of the current situation of water scarcity in Korea.

Author Contributions: H.-K.L. conceived and designed the experiments and wrote the paper.

Acknowledgments: This work was supported by the National Research Foundation of Korea (NRF) Grant funded by the Korean Government (MSIP) (No. NRF-2015R1A5A7037825).

Conflicts of Interest: The author declares no conflict of interest.

References

1. Gorb, P.; Dumas, A. Silent design. *Des. Stud.* **1987**, *8*, 150–156. [CrossRef]
2. Yudelson, J. *Dry Run: Preventing the Next Urban Water Crisis*; New Society Publisher: Gabriola Island, BC, Canada, 2010.
3. UNEP (The United Nations Environment Programme as part of its Green Economy Initiative). *Overview of The Republic of Korea's National Strategy for Green Growth*; UNEP: Nairobi, Kenya, 2010.
4. Kim, B.C. Eutrophication Status and countermeasures of the river. *Water J.* **2009**. Available online: http://www.waterjournal.co.kr/news/articleView.html?idxno=8585 (accessed on 4 November 2015).
5. Water, K. Saving Water. Available online: http://www.kwater.or.kr/info/sub01/watersavePage.do?s_mid=93 (accessed on 30 October 2015).
6. Cho, S.H.; Kang, H.J.; Park, H.C.; Rhee, E.K. A Study on the Methodology of Water Saving in Multi-Family Residential Building. *J. Korean Soc. Living Environ. Syst.* **2012**, *19*, 525–535.
7. Dragert, J. Fighting the urine blindness to provide more sanitation options. *Water SA-Pretoria* **1998**, *24*, 157–164.
8. Smith, P.G.; Reinertsen, D.G. *Developing Products in Half the Time*; Van Nostrand Reinhold: New York, NY, USA, 1991.
9. Kelley, T.; Littman, J. *The Ten Faces of Innovation*; Doubleday: New York, NY, USA, 2005.
10. Von Stamm, B. *Managing Design, Innovation and Creativity*; Wiley: Chichester, UK, 2003.
11. Keikonen, T.K.; Jääskö, V.; Mattelmäki, T.M. Three-in-one user study for focused collaboration. *Int. J. Des.* **2008**, *2*, 1–10.
12. Dosi, G. Technological paradigms and technological trajectories: A suggested interpretation of the determinants and directions of technical change. *Res. Policy* **1987**, *11*, 147–162. [CrossRef]
13. Nelson, R.R.; Winter, S.G. In search of a useful theory of innovation. *Res. Policy* **1997**, *6*, 36–67. [CrossRef]

14. Verganti, R. Radical design and technology epiphanies: A new focus for research on design management. *J. Prod. Innov. Manag.* **2011**, *28*, 384–388. [CrossRef]

15. Davies, P. Design Council (2008) Introduction to Emerging Technology; 2008. Available online: http://www.designcouncil.org.uk/AutoPdfs/Emerging_technology.pdf (accessed on 10 October 2015).

16. Moenaert, R.; De Meyer, A.; Souder, W.E.; Deschoolmeester, D. R&D/Marketing Communication During the Fuzzy Front-End. *IEEE Trans. Eng. Manag.* **1995**, *42*, 243–259.

17. Khurana, A.; Rosenthal, S.R. Integrating the fuzzy front end of new product development. *Sloan Management Review*. 15 January 1997.

18. Krishnan, V.; Ulrich, K.T. Product development decisions: A review of the literature. *Manag. Sci.* **2001**, *47*, 1–21. [CrossRef]

19. Ulrich, K.; Eppinger, S. *Product and Design Development*, 5th ed.; McGrawHill: New York, NY, USA, 2012.

20. Von Hippel, E. *Democratizing Innovation*; MIT Press: Cambridge, MA, USA, 2005.

21. Gulliksen, J.; Goransson, B.; Boivie, I.; Blomkvist, S.; Persson, J.; Cajander, A. Key principles for user-centered systems design. *Behav. Inf. Technol.* **2003**, *22*, 397–409. [CrossRef]

22. Pruitt, J.; Adlin, T. *The Persona Lifecycle: Keeping People in Mind Throughout Product Design*; Morgan Kaufmann: San Francisco, CA, USA, 2006.

23. Schaffer, E. *Institutionalization of Usability*; Pearson Education: Boston, MA, USA, 2004.

24. Patnaik, D.; Becker, R. Need finding: The why and how of uncovering people's needs. *Des. Manag. J.* **1999**, *10*, 37–43.

25. Laurel, B. *Design Research: Methods and Perspectives*; The MIT Press: Cambridge, MA, USA, 2003.

26. Mojota, B. *Design Management: Using Design to Build Brand Value and Corporate Innovation*; Allworth Press: New York, NY, USA, 2003.

27. Krippendorff, K. On the essential contexts of artifacts or on the proposition that design is making sense (of things). *Des. Issues* **1989**, *5*, 9–39. [CrossRef]

28. Mossberg, L. A marketing approach to the tourist experience. *Scand. J. Hosp. Tour.* **2007**, *7*, 59–74. [CrossRef]

29. Pine, B.J., II; Gilmore, J.H. *The Experience Economy*; Harvard Business School Press: Boston, MA, USA, 1999.

30. Voss, C.; Zomerdijk, L. *Innovation in Experiential Services—An Empirical View*; Innovation in Services; DTI: London, UK, 2007; pp. 97–134.

31. Press, M.; Cooper, R. *Design Experience*; Ashgate Publishing Limited: Farnham, UK, 2003.

32. Schön, D. *The Reflective Practitioner*; Basic Books: NewYork, NY, USA, 1983.

33. Lawson, B. *How Designers Think*, 4th ed.; Architectural Press: Oxford, UK, 2005.

34. Loch, C.; DeMeyer, A.; Pich, M.T. *Managing the Unknown: A New Approach to Managing High Uncertainty and Risk in Projects*; John Wiley and Sons: Hoboken, NJ, USA, 2006.

35. Mayhew, D.J.; Bias, R.G. *Cost-Justifying Usability*; Morgan Kaufmann: San Francisco, CA, USA, 1994.

36. Boyle, D.; Harris, M. *The Challenge of Co-Production: HOW equal Partnership between Professionals and Piblic are Crucial to Improving Public Service*; NESTA: London, UK, 2009.

37. Ramswamy, V.; Gouillart, F. Building the Co-creating Enterprise. *Harvard Business Review* 15 October 2010. 100–109.

38. Buchenau, M.; Suri, J.F. Experience prototyping. In Proceedings of the 3rd Conference on Designing Interactive Systems: Processes, Practices, Methods, and Techniques, Brooklyn, NY, USA, 17–19 August 2000; pp. 424–433.

39. Holtzblatt, K.; Wendell, J.B.; Wood, S. *Rapid Contextual Design*; Morgan Kaufmann: San Francisco, CA, USA, 2005.

40. Burr, J.; Larsen, H. The Quality of Conventions in Participatory Innovation. *Codesign* **2010**, *6*, 121–138. [CrossRef]

Communication

Optimal Design of a Residential Photovoltaic Renewable System in South Korea

Hyunkyung Shin [1] and Zong Woo Geem [2,*]

[1] Department of Financial Mathematics, Gachon University, Seongnam 13120, Korea; hyunkyung@gachon.ac.kr
[2] Department of Energy IT, Gachon University, Seongnam 13120, Korea
* Correspondence: zwgeem@gmail.com

Received: 4 February 2019; Accepted: 14 March 2019; Published: 18 March 2019

Abstract: An optimal design model for residential photovoltaic (PV) systems in South Korea was proposed. In the optimization formulation, the objective function is composed of three costs, including the monthly electricity bill, the PV system construction cost (including the government's subsidy), and the PV system maintenance cost. Here, because the monthly electricity bill is not differentiable (it is a stepped piecewise linear function), it cannot be solved by using traditional gradient-based approaches. For details considering the residential electric consumption in a typical Korean household, consumption was broken down into four types (year-round electric appliances, seasonal electric appliances, lighting appliances, and stand-by power). For details considering the degree of PV generation, a monthly generation dataset with different PV tilt angles was analyzed. The optimal design model was able to obtain a global design solution (PV tilt angle and PV size) without being trapped in local optima. We hope that this kind of practical approach will be more frequently applied to real-world designs in residential PV systems in South Korea and other countries.

Keywords: optimal design; photovoltaic system; renewables; residential building; South Korea

---

1. Introduction

South Korea is in the world's top 10 energy-consuming countries, and it heavily depends on imports of fossil fuels (natural gas, coal, and oil) [1–3]. Due to recent public awareness regarding the issue of polluted air, pressure to reduce its dependency on fossil fuels has increased. In addition, the Fukushima disaster that occurred in Japan has caused the present government to support the nuclear phase-out policy.

Therefore, various renewable energies (photovoltaics (PV), wind, geothermal, hydro, biomass, fuel cells, etc.) have been currently developed, which also helps in the country's pledge at the 2015 Paris Climate Conference to cut its carbon emissions by 37% below the business-as-usual (BAU) level by 2030.

In addition, the Korean government has recently declared a national project aiming for power generation by renewable energies to account for 20% of the total generation output by 2030 (85,905 GWh (13.6%) by 2025 and 134,136 GWh (20%) by 2030) [4]. This project especially focuses on PV and wind energies (more than 75% with respect to the generation capacity, and more than 50% with respect to the generation amount). The Korean government plans to provide urban-type self-sufficient PV systems to 760,000 residential houses by 2022, and 1,560,000 houses by 2030 [5].

To this end, as one of practical efforts, Korean government already ruled that 5% of total construction cost should be invested in renewable energy system for large public buildings (total floor area is greater than or equal to 3000 m^2), and it also subsidizes 60% of the construction cost if private residential buildings install PV renewable systems [6].

Korean government also plans to promote rural-area PV systems using low-interest loans and higher-weighted RECs (Renewable Energy Certificate). The REC is a market-tradable and non-tangible instrument that certifies that the owner possesses one megawatt-hour (MWh) of electricity generated from any renewable energy resource [7]. RPS (Renewable Portfolio Standard) required for large power producers ($\geq$500 MW) also works well after FIT (Feed-In Tariff) system ends. In order to enhance the social receptivity to PV systems, the Korean government has approved private enterprisers, to gather individual private investors, to join in PV development projects.

The objective of this study is to propose an optimal model for residential PV system design. In this model, the construction and management costs will be minimized, while considering various practical design factors such as PV generation amounts with different tilt angles, the Korean progressive electric rate, the unit cost of a PV panel, the interest rate, the project period, the electrical usage of general electric appliances, and seasonal appliances, lighting appliances, and stand-by power.

The rest of this paper is organized as follows. The optimal design model for the residential PV system is proposed in Section 2. Residential electricity demand is broken down in detail, and the monthly electrical generation amounts with varying tilt angles are proposed in the form of polynomial functions in Section 3. The optimal design solution is obtained by using an evolutionary algorithm, and compared with that from previous gradient-based methods in Section 4. Finally, in Section 5, we conclude our paper with some future directions.

2. Optimization Formulation

The objective function to be minimized in this residential PV design optimization is the total cost (C_T), which consists of the electric bill from grid ($C_{Electric}$), the PV-related construction cost (C_{Cst}), and the PV-related maintenance cost (C_{Mtn}), as shown in Equation (1) [6]:

$$\text{Minimize}\, C_T = C_{Electric} + C_{Cst} + C_{Mtn} \tag{1}$$

where the annual electric bill ($C_{Electric}$) is the sum of the monthly bills, and each monthly bill ($C_{Eelctric}^m$) is calculated based on the monthly grid-supplied amount ($D_{Electric}^m - PV_{Electric}^m$) when monthly residential demand ($D_{Electric}^m$) is greater than the monthly PV generation amount ($PV_{Electric}^m$), as in Equation (2):

$$C_{Electric} = \sum_{m=1}^{12} C_{Electric}^m (D_{Electric}^m - PV_{Electric}^m) \tag{2}$$

For the monthly bill ($C_{Electric}^m$), Korea adopts a six-stage progressive electric rate system, which charges a higher rate for higher electricity usage, as shown in Table 1.

Table 1. Korean progressive electric rate (US\$1 $\approx$ 1100 KRW).

Range	Base Rate (KRW)	Progressive Rate (KRW/kWh)
Up to 100 kWh	370	55.1
101~200 kWh	820	113.8
201~300 kWh	1430	168.3
301~400 kWh	3420	248.6
401~500 kWh	6410	366.4
More than 500 kWh	11,750	643.9

For example, if one household consumes 50 kWh for a certain month, the monthly electric bill will be 3125 KRW (=370 + 50 × 55.1); if it consumes 150 kWh, the monthly electric bill will be 12,020 KRW (=820 + 100 × 55.1 + 50 × 113.8). Thus, if we draw a monthly electric bill from 0 to 600 kWh, we obtain a stepped piecewise linear function, as shown in Figure 1.

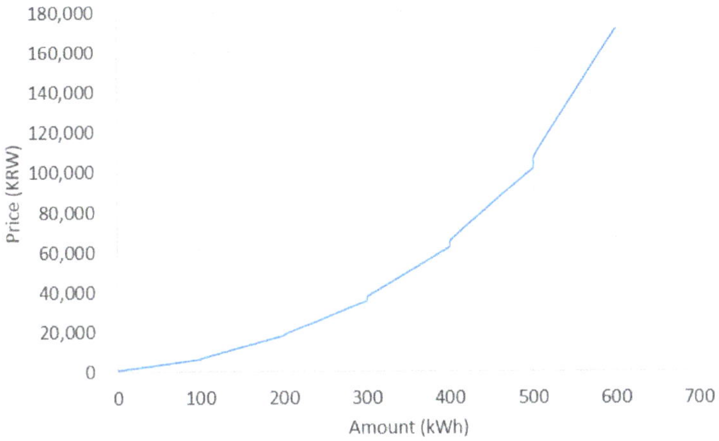

Figure 1. The six-stage progressive electric rate in Korea.

For the PV-related construction cost (C_{Cst}), in order to fairly consider this one-time cost alongside other annual costs ($C^m_{Electric}$ and C_{Mtn}), a capital recovery factor [8], which is the ratio of a constant annual return amount to the initial construction cost (C_{Icc}) for a given length of time, is introduced as in Equation (3):

$$C_{Cst} = \frac{r(1+r)^n}{(1+r)^n - 1} C_{Icc}$$ (3)

where r is the interest rate (6.5% in this study) and n is number of system operation years (or the number of annual returns received; 25 years in this study).

The decision variables in this residential PV design optimization are the size of the PV panel (or module; S^{PV}) and the tilt angle of the PV panel (A^{PV}; horizontal line is $0°$). These two decision variables have value ranges as constraints:

$$0 \leq S^{PV} \leq 3(\text{kW})$$ (4)

$$15° \leq A^{PV} \leq 60°$$ (5)

3. Application of the Residential PV System

The above formulated PV design model is assumed to be applied to a typical Korean residential building. For a typical Korean residential building, the monthly demand ($D^m_{Electric}$) can be assessed in four groups of consumption (general electric appliances, seasonal electric appliances, lighting appliances, and stand-by power) [6].

The first group of consumption occurs in general (year-round) electric appliances such as the television, refrigerator, and washing machine, as shown in Table 2. For example, a typical Korean residential building has two TV sets, which consume 270 W (=135 W × 2) over 6.9 hr per day, and 28 days per month, based on a statistical survey. Interestingly, a Korean house also possesses a special refrigerator which preserves only Kimchi, because it is an essential dish for every meal in Korean daily life.

Table 2. Power consumption of general electrical appliances.

Appliance	Power Consumption (W)	Daily Usage Hours (hr)	Monthly Usage Days (days)
Two TV sets	270	6.9	28.0
Refrigerator	67	24.0	30.0
Refrigerator for Kimchi	30	24.0	30.0
Washing Machine	515	1.5	17.5
Vacuum Cleaner	899.1	0.6	21.6
Personal Computer	168	4.2	24.4
Microwave	1010.2	0.4	14.9
Audio System	40	3.0	8.5

The second group of consumption occurs with seasonal electric appliances, such as the electric fan, air conditioner, humidifier, and electric blanket, as shown in Tables 3 and 4. For example, a typical Korean residential building has one air conditioner, which consumes 1725 W over 4.65 hr per day. However, this seasonal appliance is utilized only during the summer season (13 days for June, 15 days for July, and 27 days for August).

Table 3. Power consumption of seasonal electrical appliances.

Appliance	Power Consumption (W)	Daily Usage Hours (hr)	Yearly Usage Days (days)
Electric Fan	60	7.20	95
Air Conditioner	1725	4.65	55
Humidifier	99	5.12	126
Electric Blanket	230	5.42	146

Table 4. Monthly usage of seasonal electrical appliances.

Appliance	Days of Use in Each Month											
	1	2	3	4	5	6	7	8	9	10	11	12
Electric Fan	0	0	0	0	14	16	23	25	16	0	0	0
Air Conditioner	0	0	0	0	0	13	15	27	0	0	0	0
Humidifier	23	21	17	0	0	0	0	0	0	17	23	24
Electric Blanket	27	25	20	0	0	0	0	0	0	20	24	27

The third group of consumption occurs with lighting appliances, such as fluorescent, incandescent, and halogen lights, as shown in Table 5. For example, a typical Korean residential building has one stand-alone (stabilizer-included) fluorescent lamp, which consumes 25.86 W over 7.9 hr per day.

Table 5. Power consumption of lighting appliances.

Appliance	Power Consumption (W)	Daily Usage Hours (hr)
Fluorescent Tube (20 W)	20	7.9
Fluorescent Tube (32 W)	32.09	7.9
Fluorescent Tube (40 W)	40.18	7.9
Fluorescent (Compact)	37.90	8.1
Fluorescent (Circular)	39.85	5.6
Fluorescent (Stand-Alone)	25.86	7.9
Incandescent	71.48	1.7
Halogen	94.17	1.3

The final group of consumption occurs with stand-by power from various appliances, as shown in Table 6. Normally it accounts for approximately 10% of total household power consumption.

Table 6. Consumption amount of stand-by power.

Appliance	Average Stand-By Power (W)	Daily Stand-By Power Amount (Wh)
Two TV sets	8.6	147.1
Audio System	9.1	191.1
DVD	12.2	269.6
Microwave	2.8	66.2
Air Conditioner	2.8	54.2
Personal Computer	3.2	63.4
Computer Monitor	2.6	51.5

If we aggregate the above-mentioned four types of consumption, we can obtain a monthly power consumption graph, as shown in Figure 2. Here, it should be noted that a consumption amount of 2.2 kWh/day for any additional appliance was added to each monthly amount.

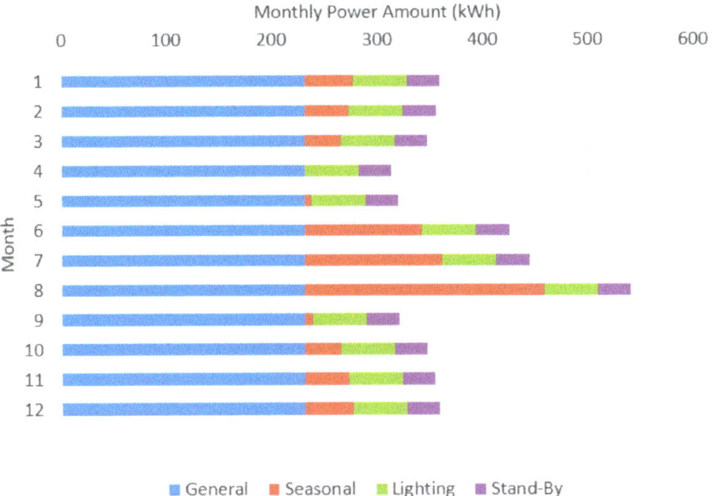

Figure 2. Monthly power consumption for a typical Korean house.

So far, the monthly power consumption of a typical Korean house has been assessed based on four different types of consumption. Now let us assess the monthly power generation amount from the PV system ($PV^m_{Electric}$).

The monthly PV generation amount is affected by two major decision variables (PV angle, A^{PV}, and PV size, S^{PV}). The first affecting factor is the tilted angle of the PV panel, as shown in Figure 3. As seen in the figure, the lowest angle (15°) generates the highest amount in June, while the highest angle (60°) generates the highest amount in December. The highest amounts in March and September occur in the middle.

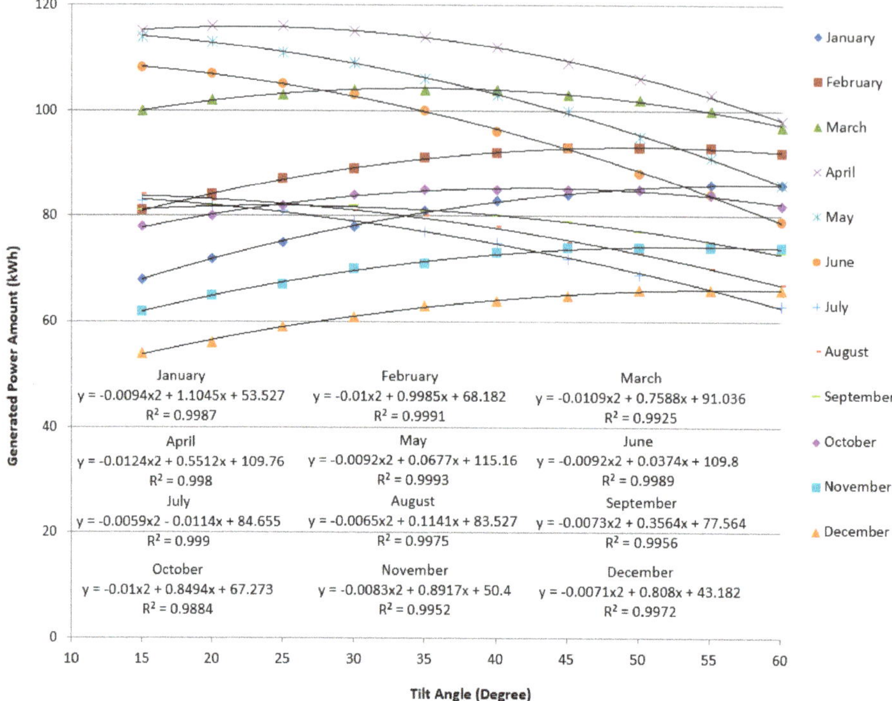

Figure 3. Monthly photovoltaics (PV) generation amounts with different tilt angles.

For this study, in order to estimate the energy production of the residential PV system, the PVWatts calculator [9], which was developed by the National Renewable Energy Laboratory (NREL) in the U.S. Department of Energy, was utilized. After inputting various PV system specifications such as the DC system size (unit size (1 kW) in this study), array type (fixed in this study), array azimuth (180° (full south) in this study), system losses (14% in this study), inverter efficiency (96% in this study), and PV tilt angle (A^{PV}) into the software, we could obtain an estimation of the month-average solar radiation (kWh/m^2/day), and the monthly unit-size PV generation amount (kWh) for a specific location.

For the specific location, this study selected Seoul, the capital city of South Korea. However, PVWatts provided the PV generation data of Incheon, as the nearest location from Seoul (24 miles west from the center of Seoul), whose latitude is 37.48° N and longitude is 126.55° E, as shown in Figure 4.

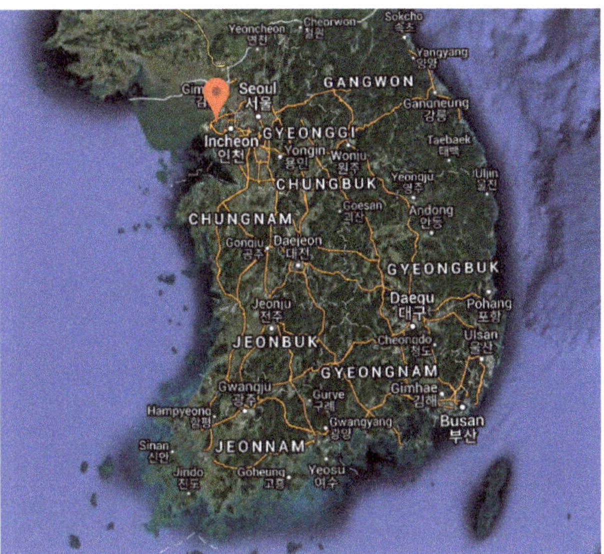

Figure 4. Location of the solar data source (Incheon) from Google Maps.

The influencing factor, PV size (S^{PV}), can be multiplied by the unit-size generation amount (kWh/kW) at a certain PV angle (A^{PV}), to calculate the monthly PV generation amount ($PV^m_{Electric}$).

The PV-related construction cost (C_{Cst}) in Equation (1) is the function of PV size (S^{PV}). The original PV construction cost is 7,210,000 KRW/kW in this study. However, after considering the Korean government's subsidy (60% of the original cost = 4,326,000 KRW/kW) and the building materials cost savings ($462,500/kW), the PV-related construction cost (C_{Cst}) becomes 2,421,500 KRW/kW (=7,210,000 − 4,326,000 − 462,500) multiplied by the PV size (S^{PV}).

The PV-related annual maintenance cost (C_{Mtn}) in Equation (1) is 12,105.7 KRW/kW (0.5% unit C_{Cst}) multiplied by the PV size (S^{PV}).

4. Computational Results

The residential PV design model is optimized with various practical data, as proposed in the above sections. Figures 5 and 6 show the total PV design cost, as specified in Equation (1), with different PV sizes ($0 \leq S^{PV} \leq 3$ kW, by 0.2 kW) and tilt angles ($15° \leq A^{PV} \leq 60°$, by 2.5°). In this resolution, 639,919 KRW, with a PV size of 1.2 kW and a PV tilt angle of 27.5° is the minimal design solution for the system.

When we narrowed down the PV size ($0.95 \leq S^{PV} \leq 1.3$ kW) and the tilt angle ($26.4° \leq A^{PV} \leq 28.6°$), and then divided them into finer intervals (0.05 kW for the PV size and 0.1° for the tilt angle), Figures 7 and 8 were obtained. At this resolution, we obtained a better solution (639,901 KRW with PV size of 1.2 kW and PV tilt angle of 28.3°~28.4°) than that with coarse resolution (639,919 KRW with PV size of 1.2 kW and PV tilt angle of 27.5°).

Size (kW)	Angle (°)																		
	15	17.5	20	22.5	25	27.5	30	32.5	35	37.5	40	42.5	45	47.5	50	52.5	55	57.5	60
0	717,546	717,546	717,546	717,546	717,546	717,546	717,546	717,546	717,546	717,546	717,546	717,546	717,546	717,546	717,546	717,546	717,546	717,546	717,546
0.2	694,407	694,068	693,802	693,608	693,487	693,438	693,462	693,558	693,727	693,968	694,282	694,668	695,130	695,693	696,330	699,032	699,818	700,679	701,613
0.4	676,073	675,374	674,811	674,386	674,097	673,944	673,929	674,050	674,308	674,702	675,234	675,902	676,707	677,648	678,727	679,942	681,293	682,782	684,407
0.6	655,356	654,280	653,440	652,774	652,282	651,964	651,820	651,849	652,052	650,439	650,990	651,720	652,712	656,881	658,233	659,758	661,457	663,375	671,050
0.8	650,905	649,663	648,643	645,858	645,366	643,126	643,114	643,356	643,807	644,468	645,339	646,418	647,708	649,206	650,915	652,832	654,959	657,296	659,841
1	645,787	644,538	643,544	642,803	640,325	640,115	640,214	640,559	641,153	642,038	643,787	645,182	646,831	648,735	650,893	653,306	655,974	658,896	662,073
1.2	644,412	642,952	641,770	640,864	640,235	639,919	641,952	642,284	642,903	643,810	645,005	646,488	648,284	651,081	653,564	656,342	659,415	662,784	666,449
1.4	649,777	648,075	646,695	645,638	644,905	644,494	644,407	644,642	645,201	646,082	647,286	648,814	650,664	652,838	655,411	660,475	663,885	667,631	671,713
1.6	655,106	653,274	651,792	650,659	649,875	649,441	649,356	649,620	650,233	651,196	652,509	654,170	656,181	658,541	661,250	664,309	667,754	672,223	676,439
1.8	659,033	656,847	655,027	653,571	651,333	650,833	652,761	653,050	653,690	654,682	659,183	661,101	663,387	666,041	669,064	672,456	676,216	680,344	684,841
2	669,794	667,668	665,888	664,455	663,369	662,630	662,812	662,305	662,883	663,822	665,108	666,741	668,722	673,245	676,680	680,161	684,137	691,727	696,724
2.2	684,973	682,556	680,506	678,364	677,246	676,482	676,072	676,137	676,701	677,627	678,907	680,543	682,585	685,633	688,601	691,938	695,645	701,895	706,652
2.4	700,831	698,455	695,872	694,443	692,950	692,331	692,066	692,153	692,593	693,386	694,532	696,030	698,614	701,622	704,567	707,900	711,543	715,991	721,451
2.6	717,795	715,595	713,778	712,344	711,292	710,621	710,334	709,891	710,484	711,447	712,779	714,482	716,554	718,996	721,807	725,012	729,510	733,784	738,444
2.8	736,695	734,433	732,543	730,527	729,563	728,958	728,712	728,825	729,298	729,719	731,393	733,197	735,370	738,480	741,517	744,932	748,725	752,895	758,276
3	756,136	754,161	752,071	749,926	749,009	748,364	748,061	748,073	750,994	752,061	753,468	755,216	757,305	759,735	762,506	766,006	769,782	773,924	779,091

Figure 5. The total PV design cost.

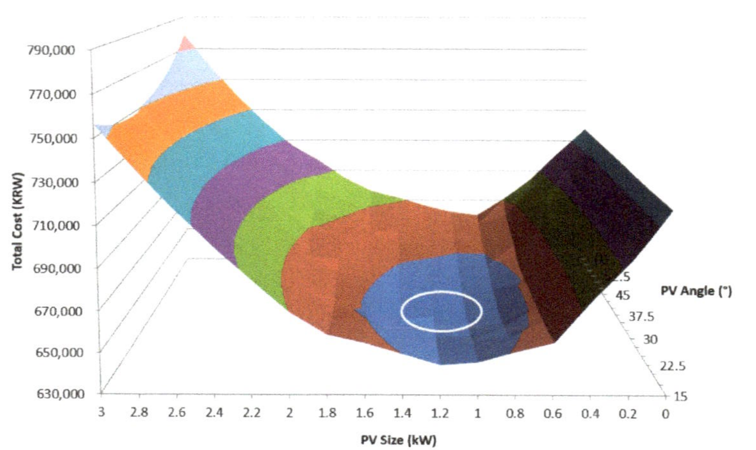

Figure 6. Map of the total PV design cost.

Size (kW)	Angle (°)																						
	26.4	26.5	26.6	26.7	26.8	26.9	27	27.1	27.2	27.3	27.4	27.5	27.6	27.7	27.8	27.9	28	28.1	28.2	28.3	28.4	28.5	28.6
0.95	642,782	642,775	642,768	642,761	642,754	642,748	642,743	642,738	642,733	642,728	642,724	642,721	642,717	642,715	642,712	642,710	642,709	642,707	642,706	642,706	642,706	642,706	642,707
1	640,162	640,153	640,145	640,138	640,131	640,128	640,125	640,122	640,119	640,118	640,116	640,115	640,114	640,114	640,114	640,114	640,115	640,116	640,118	640,120	640,122	640,125	640,128
1.05	640,587	640,581	640,576	640,572	640,568	640,564	640,561	640,558	640,555	640,553	640,552	640,551	640,550	640,549	640,549	640,550	640,551	640,552	640,554	640,556	640,558	640,561	640,564
1.1	640,504	640,496	640,488	640,481	640,474	640,467	640,461	640,456	640,450	640,446	640,441	640,438	640,434	640,431	640,428	640,426	640,424	640,423	640,422	640,422	640,421	640,422	640,423
1.15	641,242	641,234	641,226	641,218	641,211	641,204	641,198	641,192	641,187	641,182	641,177	641,173	641,170	641,166	641,164	641,161	641,159	641,158	641,157	641,156	641,156	641,157	641,157
1.2	640,004	639,991	639,978	639,966	639,958	639,951	639,945	639,939	639,933	639,928	639,923	639,919	639,915	639,912	639,909	639,906	639,905	639,903	639,902	639,901	639,901	641,892	641,892
1.25	641,162	641,148	641,135	641,122	641,110	641,098	641,086	641,075	641,065	641,055	641,045	641,036	641,027	641,019	641,011	641,004	640,997	640,991	640,985	640,980	640,975	640,970	640,966
1.3	642,320	642,305	642,291	642,278	642,265	642,253	642,241	642,230	642,219	642,208	642,198	642,189	642,180	642,171	642,163	642,156	642,148	642,142	642,136	642,130	642,125	642,120	642,116

Figure 7. The total PV design cost at a finer resolution.

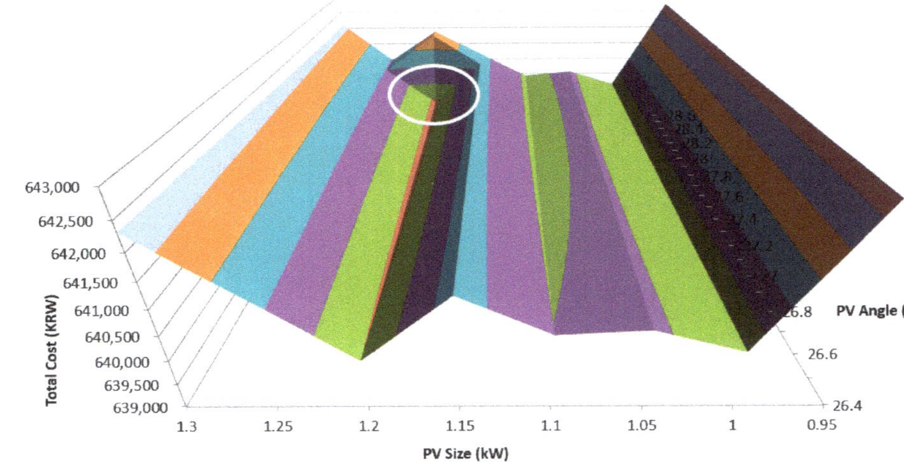

Figure 8. Map of the total PV design cost at a finer resolution.

In order to find a global optimal solution, we applied a genetic algorithm [10] as a global search meta-heuristic algorithm [11] to this PV design problem. When this meta-heuristic optimization algorithm was applied, we obtained an even better solution (639,824 KRW) at different solution spot (a PV size of 1.1904 kW and a PV tilt angle of 26.7013°) than those at the previous two resolutions. This phenomenon means that there exist local optimal solutions within the solution space.

Here, it should be noted that this PV design problem cannot be solved by using calculus-based approaches, because the monthly electric bill, as a part of the objective function, possesses stepped piecewise linearity, as shown in Figure 1. At certain stepped points such as 100, 200, 300, 400, and 500 kW, this cost function is not differentiable. Although a previous research [6] tackled this problem with a gradient-based approach, named SQP (Sequential Quadratic Programming), it had to sacrifice the accuracy of the objective function by smoothing out this step function with polynomial curve fitting.

In order to compare our approach by using a genetic algorithm with the old approach, using SQP, we first performed a polynomial regression based on the electrical rate data in Table 1, and obtained the following second-order polynomial function:

$$C_{Electric}^{m} = 0.5632x^2 - 76.207x + 7612.3 \, with \, R^2 = 0.9947 \tag{6}$$

Then, based on Equation (6), the SQP optimization was performed, obtaining an optimal cost of 617,529 KRW, with a PV size of 1.3935 and a PV tilt angle of 29.7229°. It appears that the solution (617,529 KRW) from SQP was better than that (639,824 KRW) of our approach. However, when we verified the SQP solution with a real cost table (Table 1), we obtained 644,252 KRW, which is worse than our solution.

5. Conclusions

This study proposes a design optimization model for the residential PV systems in South Korea, where the objective function to be minimized consists of three costs, such as the monthly electric bill, the PV-related construction costs, and the PV-related maintenance cost. Here, the monthly electric bill has six ranges in the form of a stepped piecewise linear function. The PV-related construction costs also include the government's subsidy and the building-material cost savings. The initial construction costs, and the annually occurring maintenance costs are fairly compared by introducing the capital recovery factor.

Regarding residential electrical consumption, four consumption types, such as year-round electric appliances, seasonal electric appliances, lighting appliances, and stand-by power, were considered. Also, regarding residential PV generation, the monthly generation amount was calculated by considering different solar altitude angles.

While local optimal solutions, this model could find the global optimal solution by using a genetic algorithm. We hope that this optimization model will be practically used in residential PV system designs in South Korea.

For future study, we plan to construct more detailed PV design optimization models by considering discrete PV size variables [12–14], ESS (energy storage systems) [15,16], AC–DC conversion [17], and more energy-efficient lighting devices (light-emitting diodes). Normally, the size of PV is discrete, because a PV system consists of an integer number of panels. Thus, we would like to consider this discrete nature of the PV size after gathering sufficient data in the future. In order to efficiently utilize surplus energy from the PV system, we may install an ESS and optimally schedule it [15].

The climate change cast over in Korea has made its summers hotter than before, which has led to more energy consumption in the summer months, and higher energy bills. Thus, the Korean government is about to reform the multi-stage progressive electric rate, in order for lower-income groups to be able to afford to pay it. Once all-new data, including billing, panel capacity and costs,

ESS capacity & costs, etc., are obtained, we will correspondingly construct a more detailed and up-to-date model design.

Author Contributions: H.S. and Z.W.G. wrote the paper, and Z.W.G. supervised the submission.

Funding: This work was supported by the Korea Institute of Energy Technology Evaluation and Planning (KETEP), and the Ministry of Trade, Industry & Energy (MOTIE) of the Republic of Korea (No. 20163010140690).

Conflicts of Interest: The authors declare no conflict of interest.

References

1. EIA (U.S. Energy Information Administration). Country Analysis Brief: South Korea. Available online: https://www.eia.gov/beta/international/analysis.php?iso=KOR (accessed on 4 February 2019).
2. Geem, Z.W.; Roper, W.E. Energy Demand Estimation of South Korea Using Artificial Neural Network. *Energy Policy* **2009**, *37*, 4049–4054. [CrossRef]
3. Geem, Z.W.; Kim, J.-H. Optimal Energy Mix with Renewable Portfolio Standards in Korea. *Sustainability* **2016**, *8*, 423. [CrossRef]
4. Kang, K.-S.; Kim, M.-S.; Kwak, J.-Y. Renewable Energy Scenarios for 2030 in Korea. *J. Wind Energy* **2017**, *8*, 5–10.
5. MOTIE (Ministry of Trade, Industry and Energy). *White Paper on Trade, Industry and Energy (Energy Section)*; Korean Ministry of Trade, Industry and Energy: Sejong, Korea, 2018.
6. Jeon, J.-P.; Kim, K.-H. An Optimal Decision Model for Capacity and Inclining Angle of Residential Photovoltaic Systems. *Trans. Korean Inst. Electr. Eng.* **2010**, *59*, 1046–1052.
7. Renewable Energy Certificate. Available online: https://en.wikipedia.org/wiki/Renewable_Energy_Certificate_(United_States) (accessed on 4 February 2019).
8. Mays, L.W.; Tung, Y.K. *Hydrosystems Engineering and Management*; McGraw-Hill: New York, NY, USA, 1992.
9. NREL (The National Renewable Energy Laboratory). PVWatts® Calculator. Available online: https://pvwatts.nrel.gov/ (accessed on 4 February 2019).
10. Goldberg, D.E. *Genetic Algorithms in Search Optimization and Machine Learning*; Addison-Wesley: Reading, MA, USA, 1989.
11. Saka, M.P.; Hasançebi, O.; Geem, Z.W. Metaheuristics in Structural Optimization and Discussions on Harmony Search Algorithm. *Swarm Evol. Comput.* **2016**, *28*, 88–97. [CrossRef]
12. Geem, Z.W. Size Optimization for a Hybrid Photovoltaic-Wind Energy System. *Int. J. Electr. Power Energy Syst.* **2012**, *42*, 448–451. [CrossRef]
13. Askarzadeh, A. A discrete chaotic harmony search-based simulated annealing algorithm for optimum design of PV/wind hybrid system. *Sol. Energy* **2013**, *97*, 93–101. [CrossRef]
14. Askarzadeh, A. Developing a discrete harmony search algorithm for size optimization of wind–photovoltaic hybrid energy system. *Sol. Energy* **2013**, *98*, 190–195. [CrossRef]
15. Geem, Z.W.; Yoon, Y. Harmony Search Optimization of Renewable Energy Charging with Energy Storage System. *Int. J. Electr. Power Energy Syst.* **2017**, *86*, 120–126. [CrossRef]
16. Aryani, D.R.; Kim, J.-S.; Song, H. Suppression of PV Output Fluctuation Using a Battery Energy Storage System with Model Predictive Control. *Int. J. Fuzzy Logic Intell. Syst.* **2017**, *17*, 202–209. [CrossRef]
17. Fitri, I.R.; Kim, J.-S. An Optimal Current Control of Interlink Converter Using an Explicit Model Predictive Control. *Int. J. Fuzzy Logic Intell. Syst.* **2018**, *18*, 284–291. [CrossRef]

Article

Eco-Friendly Education Facilities: The Case of a Public Education Building in South Korea

Eunil Park [1] and Angel P. del Pobil [1,2,*]

[1] Department of Interaction Science, Sungkyunkwan University, Seoul 03063, Korea; pa1324@gmail.com or eunilpark@skku.edu

[2] Department of Computer Science and Engineering, Jaume-I University, 12071 Castellón, Spain

* Correspondence: pobil@uji.es; Tel.: +34-964-72-82-93

Received: 23 August 2018; Accepted: 20 September 2018; Published: 25 September 2018

Abstract: Since the importance and effects of national energy policies, plans, and roadmaps were presented in South Korea, the role of renewable energy resources has received great attention. Moreover, as there is significant reasoning for reducing and minimizing nuclear and fossil fuel usage in South Korean national energy plans, several academic scholars and implementers have expended significant effort to present the potential and feasibility of renewable energy resources in South Korea. This study contributes to these efforts by presenting potential sustainable configurations of renewable energy production facilities for a public building in South Korea. Based on economic, environmental, and technical information as well as the presented simulation results, it proposes an environmentally friendly renewable energy production facility configuration that consists of photovoltaic arrays, battery units, and a converter. Subsidies for installing and renovating such facilities are also considered. The potential configuration indicates $0.464 as the cost of energy, 100% of which is renewable. Potential limitations and future research areas are suggested based on the results of these simulations.

Keywords: eco-friendly; renewable energy; energy subsidies; SEMS; South Korea

1. Introduction

After the peaceful turnover of political power in 2017, the new South Korean government is attempting to reform national energy policies and plans. During this reform process, the government is aiming to phase out the usage of nuclear energy. One of the major decisions undertaken in this regard was the suspension of the construction of the fifth and sixth nuclear reactors in Kori when the construction process was 28.8% complete [1]. Although construction resumed when the government was implored to do so by a jury of 471 randomly selected citizens, the government decided that no new nuclear power plants in South Korea should be constructed [2].

Therefore, exploring, adopting, and using new energy resources is one of the most important research areas for establishing national energy policies and plans. As nuclear energy constitutes approximately 12% of South Korea's national energy supply system, alternative energy resources could be unpalatable or infeasible [3]. Moreover, because of the Paris Agreement, which requires South Korea to reduce its large greenhouse gas emissions, increasing the use of fossil fuels would also be infeasible [4].

Accordingly, the South Korean government is aiming to implement renewable energy resources. Approximately 4.6% of primary energy in South Korea was provided by renewable energy production facilities in 2015, with an annual growth rate of 15.2% [3]. Moreover, South Korea's energy dependence on foreign countries is very high, and compared with other countries, oil price fluctuations have a greater effect on the South Korean economy [5]. Thus, ensuring reliable energy production and securing stable energy resources are important.

The South Korean government has pursued the expansion and distribution of renewable energy production facilities to achieve its goal that, by 2035, 11% of primary energy in South Korea should be generated from renewable energy resources. To achieve this goal, the government established *The fourth basic plans for the technology development, usage, and distribution of new and renewable energy*, with detailed programs for promoting each energy source [6]. Based on these promotion programs, the duty ratio of the renewable portfolio standards was revised and enhanced. All public buildings with a certain total floor area that are planned to be constructed, reconstructed, or enlarged should have at least 21% of their energy supplied by renewable resources [7]. Therefore, both central and local South Korean governments urge public organizations and institutions to include electricity and energy production facilities that use renewable resources. For this reason, several studies have attempted to investigate the feasibility of implementing renewable energy production facilities in diverse public buildings, including a public university, local government office, and multi-purpose public buildings [8].

Considering this background, this paper presents a case study for the economic and environmental feasibility of renewable energy production systems in a public education building in South Korea. After careful simulation, a potential configuration of a renewable energy production system for the building is suggested.

Review of Prior Feasibility Studies in South Korea

Several studies have attempted to investigate the feasibility of renewable-oriented power generation facilities in South Korea. Table 1 summarizes the key prior studies. As presented in Table 1, the majority of prior studies have focused on specific regions such as certain islands or areas. This means that only a few studies have explored the sustainable cases of renewable energy facilities for public buildings in South Korea.

Table 1. Summary of prior feasibility studies of renewable energy facilities conducted in South Korea (W: Wind turbines, P: photovoltaic (PV) arrays, D: Diesel generators, B: Battery units, C: Converters, K: Kerosene generators).

Target Location	Suggested Configurations	Cost of Energy ($ per kWh)	Sources
Jeju (island)	W-P-D-B-C systems	0.174	[9]
Busan Asiad Main Stadium (sports complexes)	W-P-D-B-C systems	0.491	[10]
Hongdo (island)	K-W-D-B-C systems	0.303	[11]
Jeju World Cup Venue (sports complexes)	W-P-B-C systems	0.405	[12]
Gadeokdo (island)	W-P-B-C systems	0.326	[13]
Semiconductor facilities (industrial facilities)	P-B-C systems	0.668	[14]
Kyung-Hee University (education facilities)	W-P-B-C systems	0.509	[15]
Geojedo (island)	W-P-B-C systems	0.472	[16]
Gasado (island)	W-P-B-C systems	1.284	[17]

2. Case Study: Research Background

2.1. Location and Facilities

This study focused on an elementary school located in southeastern South Korea. To investigate the economic and environmental feasibility of potential renewable energy production facility configurations for school buildings, this study selected *Samrangjin Elementary School in Miryang* (SESM), which has smart meter facilities to record the school's hourly electricity usage. The location of this school is 35°23′44.76″ N and 128°50′15.89″ E, and it has approximately 100 students and 40 members of staff. The school has four buildings: the main education building, education support building, warehouse, and kindergarten building. Figure 1 presents the location and an overview of SESM in South Korea.

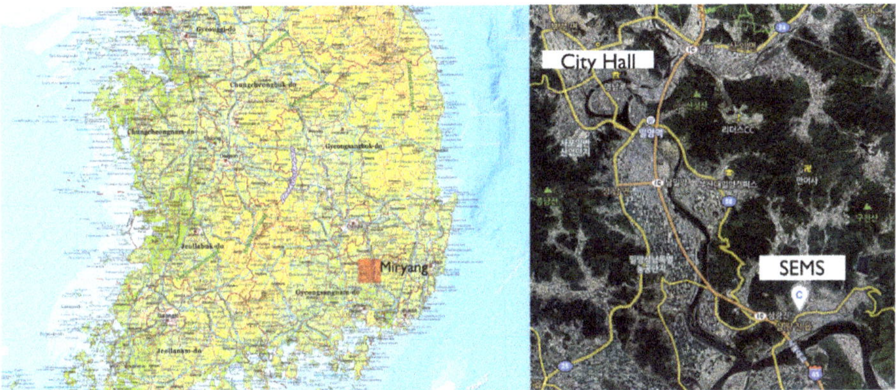

Figure 1. The location of SESM in South Korea.

2.2. Energy Load Information

SESM's energy system mainly uses electricity provided by the national grid. In 2016, SESM used 99,263 kWh. The SESM electricity load shows a scaled annual average electricity level of 223 kWh/d with a load factor of 0.464.

2.3. Renewable Energy Resources

This study uses the solar resource data provided by the National Aeronautics and Space Administration (NASA). Table 2 summarizes the monthly baseline data. Annual average solar daily radiation is 4.259 kWh/m^2/d, with a solar clearness index of 0.514. The wind resource information of SESM is collected and provided by the Korea Meteorological Administration. Figure 2 summarizes the monthly wind speed.

Table 2. Monthly solar resource information for SESM.

Month	Solar Clearness Index	Solar Daily Radiation (kWh/m^2/d)
January	0.528	2.646
February	0.513	3.255
March	0.504	4.120
April	0.547	5.428
May	0.522	5.793
June	0.498	5.744
July	0.479	5.404
August	0.499	5.159
September	0.496	4.339
October	0.539	3.706
November	0.556	2.950
December	0.547	2.515

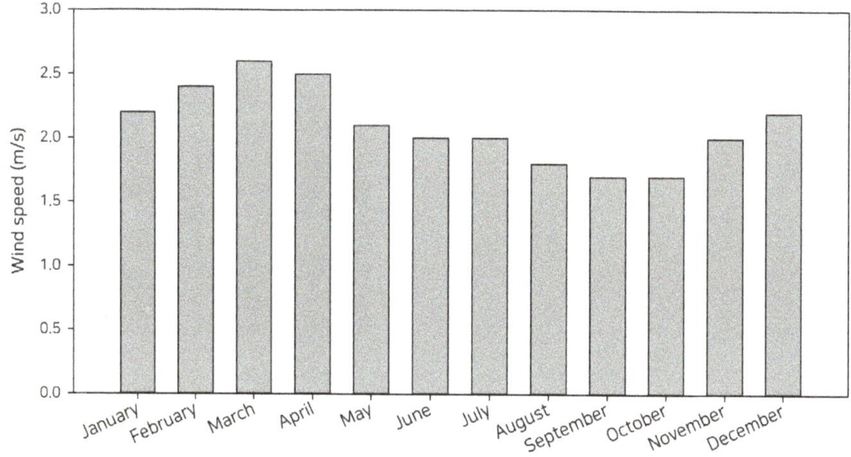

Figure 2. Average wind speeds at SESM.

3. Simulation Parameters

3.1. Annual Real Interest Rate

To present precise simulation results, the actual South Korean interest rate is computed and used in the simulations [18,19]. Based on a report by the Bank of Korea, an annual interest rate of 3.51% is used.

3.2. Evaluation Criteria

To explore the results of the simulation, the suggested sustainable configurations are ordered by two economic outputs: the cost of energy (COE) and net present cost (NPC). The COE is defined as *the mean cost of generating 1 kWh electricity by the suggested configuration*, while the NPC is defined as *the total cost of installing, utilizing, replacing, and performing the functions of the suggested configuration throughout the project* [15]. In addition, the project lifespan used by this study is 25 years.

4. Renewable Electricity Production Systems

To introduce the configurations of the sustainable renewable energy production system, the economic information of each component that can be used in the configurations should be investigated. Based on detailed economic information about the components from previous studies, Table 3 summarizes the specific economic information about the components employed in the simulations presented here. In addition, the standard electricity price, which is introduced by KEPCO, is used in the grid connection.

Table 3. Economic information about the components (* The supporting policies and plans by the South Korean government are applied).

Component	Size	Cost Information	Lifetime (Years)	Considered Capacity	Others
PV array	1.0 kW	$1500 and 750 * (Capital), $1500 and 750 * (Replacement), $25 per year (Operation & management)	20	0–600 kW	Derating factor: 80% reflectance: 20%
Wind turbine	1 unit	$1960 and 980 * (Capital), $1960 and 980 * (Replacement), $30 per year (Operation & management)	20	0–2 units	Model: Generic 1 kW turbine Hub height: 25 m
Battery (S6CS25P)	1 unit	$1229 (Capital), $1229 (Replacement), $10 per year (Operation & management)	–	0–300 units	Nominal capacity: 1156 Ah (6.94 kWh) Lifetime throughput: 9645 kWh Nominal voltage: 6 V
Converter	1.0 kW	$1000 (Capital), $1000 (Replacement), $10 per year (Operation & management)	15	0–400 kW	90% efficiency

5. Results

The sustainable configuration, which contains PV panels, battery units, and a converter, is proposed based on the simulation results (Table 4). Table 5 summarizes the total and annual costs of the proposed configuration. The suggested configuration for providing reliable and sustainable energy services to SESM includes 500 kW-capacity PV arrays, a 247 kW-capacity converter, and 202 battery units (Table 4).

Table 4. Summary of the suggested configuration from the simulation results for SEMS.

Components	Index	Components	Index
PV array	500 kW-capacity	Wind turbine	0 unit
Battery units	202 units	Converter	247 kW-capacity
Initial capital cost	$870,258	Operating cost	−$14,965 per year
Total net present cost	$623,617	Cost of energy	$0.464 per kWh
Renewable fraction	1.00		

Table 5. Costs of the suggested energy system configuration for SEMS.

Cost Category	Component ($)	Capital ($)	Replacement ($)	O&M ($)	Salvage ($)	Total ($)
Total NPC ($)	PV array	375,000	188,462	206,019	−119,010	650,471
	Grid connection	–	–	−885,431	–	−885,431
	Battery units	248,258	273,020	33,293	−96,296	458,275
	Converter	247,000	147,432	40,709	−34,839	400,302
	System	870,258	608,914	−605,410	−250,145	623,617
Annual cost ($ per year)	PV array	22,753	11,435	12,500	−7221	39,467
	Grid connection	–	–	−53,723	–	−53,723
	Battery units	15,063	16,565	2020	−5843	27,805
	Converter	14,986	8945	2470	−2114	24,288
	System	52,802	36,945	−36,733	−15,177	37,837

The estimated net present and annual costs are $623,617 and $37,837, respectively. The computed COE is $0.464 per kWh. Table 6 provides the annual electricity production. All the energy provided by the suggested system originates from renewable sources. As shown in Table 6, approximately 14% of the electricity produced by the suggested system fulfils the electricity demand of SEMS (AC primary load), while 86% is sold through the grid connection.

Table 6. Annual electricity consumption and production of the suggested configuration.

Load	Consumption (kWh Per year)	Fraction	Component (kWh Per year)	Production	Fraction
AC primary load	81,330	0.14	PV array	705,751	1.00
Grid sales	488,388	0.86	Grid purchases	0	0
Total	569,718	1.00	Total	705,751	1.00

Altering the current grid connection to the suggested configuration is expected to reduce annual emissions of carbon dioxide, sulfur dioxide, and nitrogen oxide by 308,661 kg, 1338 kg, and 654 kg, respectively.

6. Discussion and Conclusions

To develop more sustainable and eco-friendly energy plans in South Korea, the government intends to distribute renewable energy generation facilities to public buildings and organizations. Considering this trend, this study introduces a potential sustainable renewable energy generation facility configuration to fulfil the electricity demand of SEMS using local, natural resources. To evaluate the suggested configurations of the simulations, both the COE and the NPC are computed and employed.

The suggested configuration achieves 100% renewable energy, with a COE of $0.464. Although the COE of this configuration is higher than the current price of the South Korean grid connection [20], the suggested configuration can be applied to SEMS, a public education building in South Korea, as an on-site test. In addition, the simulation results also indicate that subsidies are an important issue in distributing and maintaining renewable energy production facilities [21].

As using renewable energy production facilities significantly reduces greenhouse gas emissions, the South Korean government and associated industries should distribute renewable energy production facilities [22]. According to the Paris Agreement, which introduces a mandatory level of greenhouse gas emissions for 195 nations, the South Korean government should also attempt to distribute more sustainable energy production facilities within its electricity system [23].

Although both economic and environmental information, which can be applied to the potential configuration of SEMS, was investigated considering subsidies for renewable energy production facilities in South Korea, this study has several limitations. First, the economic aspects of the certified reduction in greenhouse gas emissions were not considered. This can produce better economic results on utilizing renewable energy resources than the simulation results [24]. Second, economic theories in the renewable energy industry were not considered in the simulations. Several studies have already indicated that various economic theories can be applied and used in the renewable energy industry and market [25,26]. Therefore, future research should aim to eliminate the limitations of the current study.

Author Contributions: E.P. designed and wrote the majority of the manuscript. A.P.d.P. contributed to the analysis and revised the manuscript.

Funding: This work was supported by the Ministry of Education of the Republic of Korea and the National Research Foundation of Korea (NRF-2018S1A5A8027730). This research was also supported by the MIST, Korea, under the National Program for Excellence in SW supervised by the IITP (2015-0-00914). Moreover, the second author's research is funded by Ministerio de Economía y Competitividad (DPI2015-69041-R) and by Universitat Jaume I.

Conflicts of Interest: The authors declare no conflict of interest.

References

1. Roh, S.; Kim, D. Effect of Fukushima accident on public acceptance of nuclear energy (Fukushima accident and nuclear public acceptance). *Energy Sources Part B Econ. Plan. Policy* **2017**, *12*, 565–569. [CrossRef]
2. Ministry of Trade, Industry and Energy (South Korea). *South Korea Government, the Introduction of Re-Construction of New Kori Nuclear Plants and Energy Roadmap*; Ministry of Trade, Industry and Energy: Sejong, Korea, 2017.
3. Korea, S. *The Current Energy Status in Korea*; Ministry of Trade, Industry and Energy: Sejong, Korea, 2017.
4. Kraxner, F.; Leduc, S.; Lee, W.; Son, Y.; Kindermann, G.; Patrizio, P.; Mesfun, S.; Yowargana, P.; Mac Dowall, N.; Yamagata, Y.; et al. Towards the Paris Agreement—Negative emission and what Korea can contribute. In Proceedings of the 19th EGU General Assembly, EGU2017, Vienna, Austria, 23–28 April 2017.
5. Chung, W.S.; Kim, S.S.; Moon, K.H.; Lim, C.Y.; Yun, S.W. A conceptual framework for energy security evaluation of power sources in South Korea. *Energy* **2017**, *137*, 1066–1074. [CrossRef]
6. Hong, H.; Choi, J.; Im, S. Renewable energy production by heat pump as renewable energy equipment. *Korean J. Air-Cond. Refrig. Eng.* **2017**, *29*, 551–557.
7. Lee, C.Y.; Huh, S.Y. Forecasting the diffusion of renewable electricity considering the impact of policy and oil prices: The case of South Korea. *Appl. Energy* **2017**, *197*, 29–39. [CrossRef]
8. Oh, S.D.; Yoo, Y.; Song, J.; Song, S.J.; Jang, H.N.; Kim, K.; Kwak, H.Y. A cost-effective method for integration of new and renewable energy systems in public buildings in Korea. *Energy Build.* **2014**, *74*, 120–131. [CrossRef]
9. Kim, H.; Baek, S.; Park, E.; Chang, H.J. Optimal green energy management in Jeju, South Korea–On-grid and off-grid electrification. *Renew. Energy* **2014**, *69*, 123–133. [CrossRef]
10. Park, E.; Kwon, S.J. Renewable energy systems for sports complexes: A case study. *Proc. Inst. Civ. Eng. Energy* **2017**, *171*, 49–57. [CrossRef]
11. Bae, K.; Shim, J.H. Economic and environmental analysis of a wind-hybrid power system with desalination in Hong-do, South Korea. *Int. J. Precis. Eng. Manuf.* **2012**, *13*, 623–630. [CrossRef]
12. Park, E.; Kwon, S.J.; Del Pobil, A.P. For a Green Stadium: Economic feasibility of sustainable renewable electricity generation at the Jeju world cup venue. *Sustainability* **2016**, *8*, 969. [CrossRef]
13. Park, E.; Kwon, S.J. Towards a Sustainable Island: Independent optimal renewable power generation systems at Gadeokdo Island in South Korea. *Sustain. Cities Soc.* **2016**, *23*, 114–118. [CrossRef]
14. Choi, H.J.; Han, G.D.; Min, J.Y.; Bae, K.; Shim, J.H. Economic feasibility of a PV system for grid-connected semiconductor facilities in South Korea. *Int. J. Precis. Eng. Manuf.* **2013**, *14*, 2033–2041. [CrossRef]
15. Park, E.; Kwon, S.J. Solutions for optimizing renewable power generation systems at Kyung-Hee University's Global Campus, South Korea. *Renew. Sustain. Energy Rev.* **2016**, *58*, 439–449. [CrossRef]
16. Park, E.; Yoo, K.; Ohm, J.Y.; Kwon, S.J. Case study: Renewable electricity generation systems on Geoje Island in South Korea. *J. Renew. Sustain. Energy* **2016**, *8*, 015904. [CrossRef]
17. Chae, W.K.; Lee, H.J.; Won, J.N.; Park, J.S.; Kim, J.E. Design and field tests of an inverted based remote microgrid on a Korean Island. *Energies* **2015**, *8*, 8193–8210. [CrossRef]
18. Park, E. Potentiality of renewable resources: Economic feasibility perspectives in South Korea. *Renew. Sustain. Energy Rev.* **2017**, *79*, 61–70. [CrossRef]
19. Dursun, B. Determination of the optimum hybrid renewable power generating systems for Kavakli campus of Kirklareli University, Turkey. *Renew. Sustain. Energy Rev.* **2012**, *16*, 6183–6190. [CrossRef]
20. KEPCO. *Electric Rates Tables in Korea*; Korea Electric Power Corporation: Naju, Korea, 2017.
21. Kwon, T.H. Is the renewable portfolio standard an effective energy policy?: Early evidence from South Korea. *Util. Policy* **2015**, *36*, 46–51. [CrossRef]
22. Kafle, S.; Parajuli, R.; Bhattarai, S.; Euh, S.H.; Kim, D.H. A review on energy systems and GHG emissions reduction plan and policy of the Republic of Korea: Past, present, and future. *Renew. Sustain. Energy Rev.* **2017**, *73*, 1123–1130. [CrossRef]
23. Larkin, A.; Kuriakose, J.; Sharmina, M.; Anderson, K. What if negative emission technologies fail at scale? Implications of the Paris Agreement for big emitting nations. *Clim. Policy* **2018**, *18*, 690–714. [CrossRef]
24. Ishikawa, J.; Okubo, T. Greenhouse-Gas Emission Controls and Firm Locations in North–South Trade. *Environ. Resour. Econ.* **2017**, *67*, 637–660. [CrossRef]

25. Lesser, J.A.; Su, X. Design of an economically efficient feed-in tariff structure for renewable energy development. *Energy Policy* **2008**, *36*, 981–990. [CrossRef]
26. Traber, T.; Kemfert, C. Impacts of the German support for renewable energy on electricity prices, emissions, and firms. *Energy J.* **2009**, *30*, 155–178. [CrossRef]

MDPI

St. Alban-Anlage 66

4052 Basel

Switzerland

Tel. +41 61 683 77 34

Fax +41 61 302 89 18

www.mdpi.com

Applied Sciences Editorial Office

E-mail: applsci@mdpi.com

www.mdpi.com/journal/applsci